LibrariesWest

The Met Office Book of the British Weather

AUTHORED BY JOHN PRIOR
EDITED BY SARAH TEMPEST

Met Office
D&C
David and Charles

A DAVID & CHARLES BOOK

David & Charles is an F+W Media, Inc. company
4700 East Galbraith Road
Cincinnati, OH 45236

First published in the UK in 2010

A catalogue record for this book is available from the British Library.

ISBN-13: 978-0-7153-3640-3
ISBN-10: 0-7153-3640-1

Printed in China by RR Donnelley
for David & Charles
Brunel House, Newton Abbot, Devon

Publisher: Stephen Bateman
Acquisitions Editor: Neil Baber
Editor: Sarah Callard
Senior Designer: Martin Smith
Production Controller: Beverley Richardson

Contents

Introduction

'Climate is what we expect, weather is what we get.'
Mark Twain (1835–1910)

The UK is well known for the variability of its weather – from place to place, day to day, season to season and year to year. Its position plays a major role in this: being on the edge of the relatively warm waters of the Atlantic Ocean, yet close enough to mainland Europe to be influenced by the continental land mass. It could be said that the UK is located at a 'meteorological crossroads' since the direction from which an air stream arrives has a huge bearing on the weather – especially on temperatures and cloud cover. Sources can be as varied as the Arctic, Siberia and North Africa but predominantly it's the North Atlantic, with westerly winds moving weather systems from west to east.

Changes in land-use and altitude over relatively short distances, together with a coastline some 11,000 miles long and numerous islands, add to the weather's variability. Temperatures in a city can be warmer than those in the surrounding countryside and a seaside resort can be bathed in sunshine while the hills inland are cloaked in cloud and rain.

In general, places in the east and south of the UK tend to be drier, warmer, sunnier and less windy than those further west and north. These favourable weather conditions usually occur more often in the spring and summer than in the autumn and winter. But that is by no means the whole story, and there are intriguing local, regional and seasonal variations.

The weather: a very British obsession

The British are fascinated by the weather and are famous for their ability to talk about it at length. There is a long history of observing and recording the weather's ups and downs. This came to the fore in Victorian Britain, when weather stations were established, observing standards were set, new instruments were developed and monthly reports began to be published.

The Met Office can trace its roots back to this period as its forerunner, the 'Meteorological

Department of the Board of Trade' was set up in 1854 under Vice-Admiral Robert FitzRoy, who had earlier commanded HMS Beagle with the young naturalist Charles Darwin on board. By 1900 there were about 70 climate stations gathering a variety of daily statistics and over 3,000 rainfall stations collecting daily or monthly figures. Soon, it was possible to prepare summaries and maps of the British climate, representing averages and extremes over a long period – usually 30 years.

During the 20th century, the network of weather stations continued to grow, to take in sites at holiday resorts, reservoirs, coastguard stations, schools and research stations as well as those at airports and military bases manned by Met Office staff. Each day dedicated weather watchers made observations of temperature extremes, rainfall totals, sunshine hours, wind speed, cloud cover, snow depth and other weather characteristics. A huge archive of weather data has been built up, with almost all of it from the last 50 years in digital format. This digitised station data allows more detailed analyses of weather patterns. These include the production of sets of monthly, seasonal and annual averages, that can be plotted as colour-shaded maps.

These maps form the basis of the first five chapters of this book and present a picture of the British climate over the 30-year period, 1971 to 2000. Here, standard meteorological seasons have been used: spring is March to May, summer is June to August, autumn is September to November and winter is December to February. These detailed maps show how the climate differs from region to region.

Computer technology has also made it possible to study climate trends, to produce mathematical models of the atmosphere and use these to identify the human and natural influences on the world's climate. It has been possible to predict the climate of the future, assuming different levels of industrial activity. So the final chapter of this book looks ahead to how the British climate might change as this century progresses.

The Weather Month-by-month

The monthly maps in this chapter reveal the changing fortunes of different parts of the country as the year passes. Some features are always present, whatever the month: generally speaking it is cooler, wetter, windier and less sunny the further north and the higher up you are. However, there are some fascinating exceptions that include the Northern Ireland coast experiencing a warmer January than in the land-locked East Midlands; the Western Isles being sunnier in May than the Thames Valley, and Norfolk being wetter in July than Fife. Local microclimates add another dimension – built-up areas provide extra warmth, mountains can shelter places nearby, especially those to the east, and coasts are often favoured for sunshine.

The sea takes longer than the land to warm up in summer and cool down in the winter, helping to make July and August the warmest months and January and February the coldest.

Wind and rain are closely related, so the months of October to January tend to be both the windiest and the wettest – especially in areas closest to the Atlantic Ocean. However, in eastern England summer showers can make July and August almost as wet as the winter months. The driest weather usually occurs between February and May.

Sunshine tends to mirror day length, so it's in shortest supply in December. The sunniest time of year varies, with July usually finest over much of southern and eastern England and May the firm favourite further north and west.

Hours of Sunshine
Month-by-month

Whatever the month, if you are in the south, near the coast and at low altitude you will enjoy the most sunshine. The sunniest places of all are the seaside resorts of Sussex, Hampshire and the Isle of Wight - the highest recorded monthly total was an impressive 384 hours at Eastbourne in July 1911. To the north-west, May is the month for reaching high sunshine values - 300 hours were recorded on the Isle of Skye in May 1975.

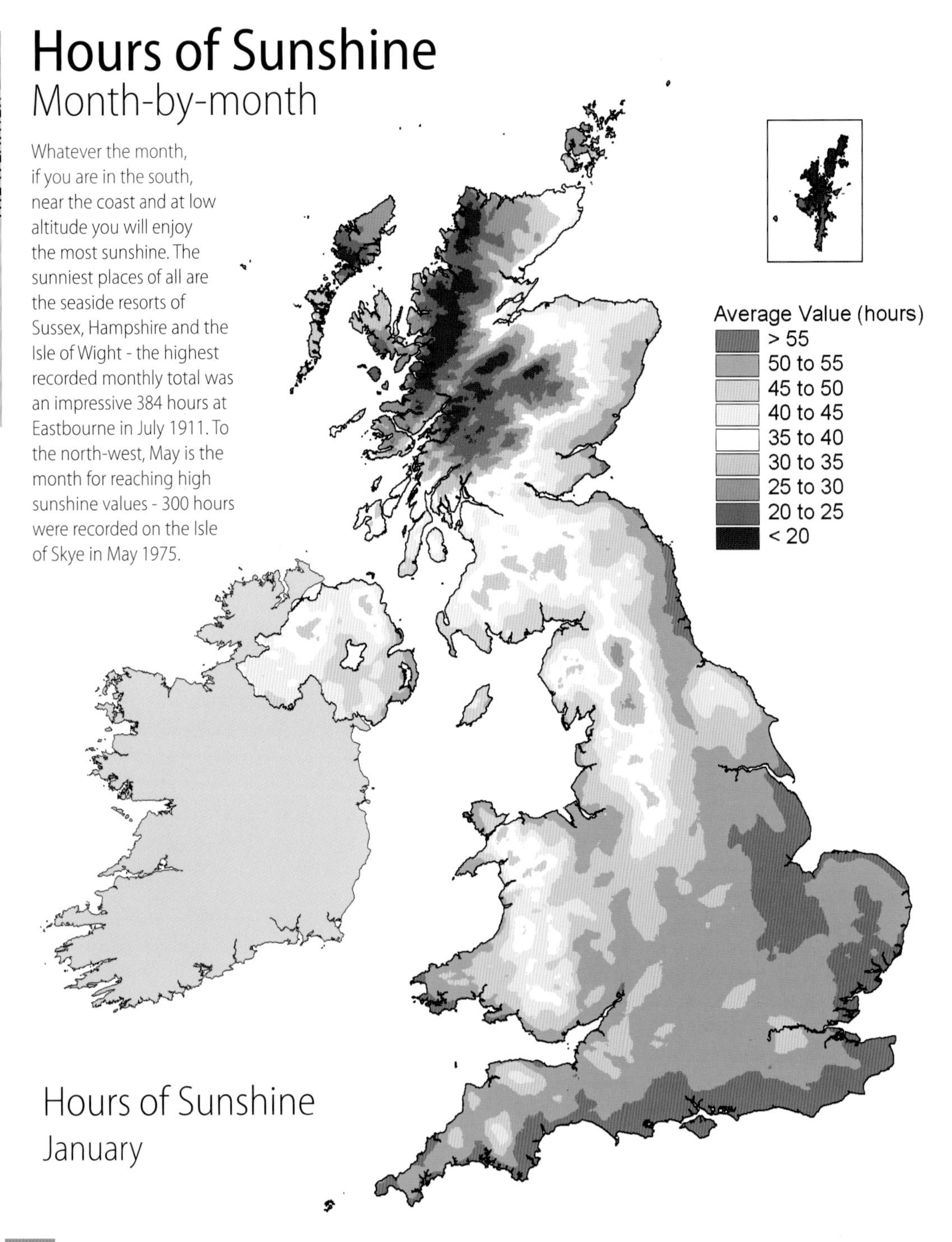
Average Value (hours)
> 55
50 to 55
45 to 50
40 to 45
35 to 40
30 to 35
25 to 30
20 to 25
< 20

Hours of Sunshine
January

Hours of Sunshine
February

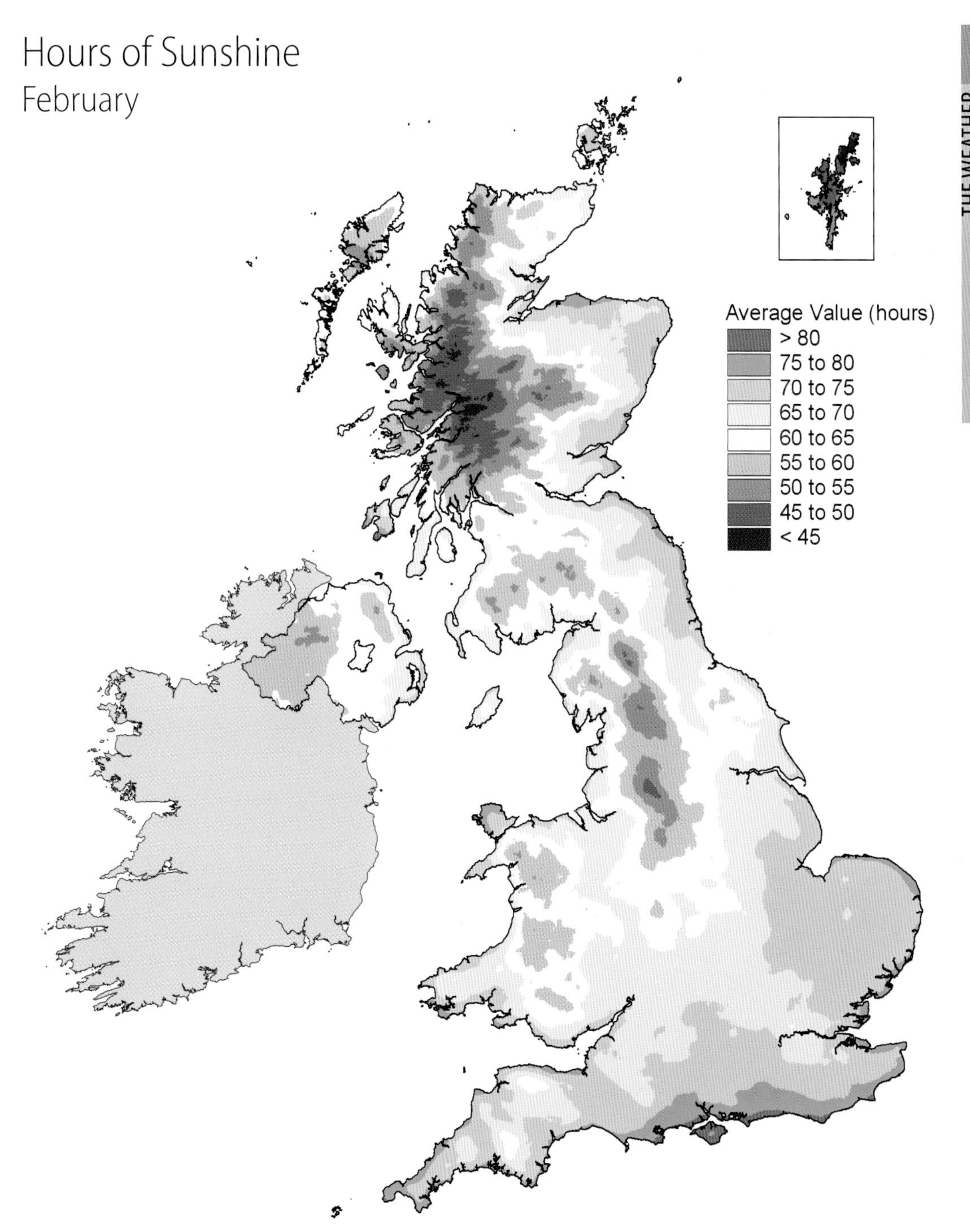

Average Value (hours)
> 80
75 to 80
70 to 75
65 to 70
60 to 65
55 to 60
50 to 55
45 to 50
< 45

Hours of Sunshine
March

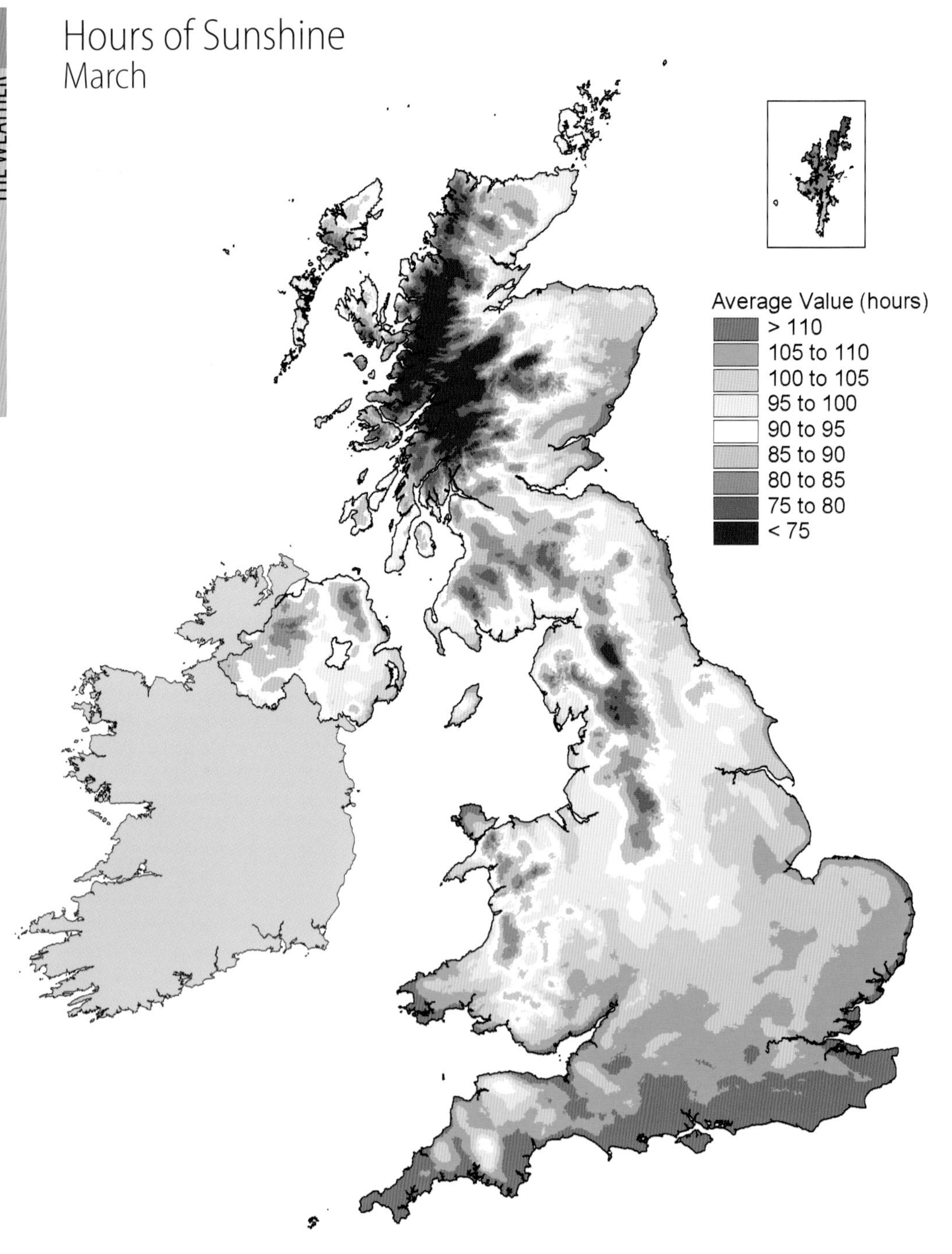
Average Value (hours)
> 110
105 to 110
100 to 105
95 to 100
90 to 95
85 to 90
80 to 85
75 to 80
< 75

Hours of Sunshine
April

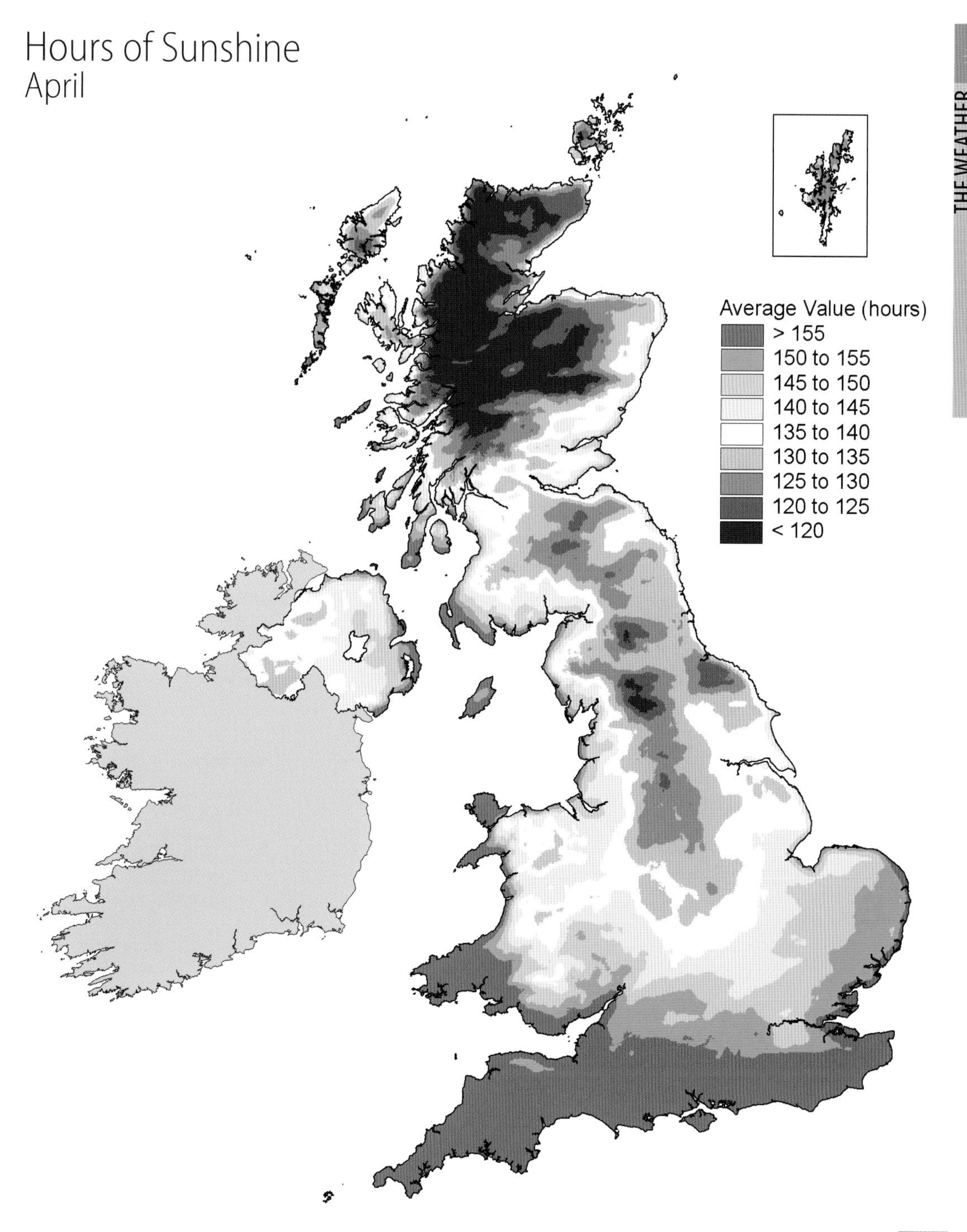
Average Value (hours)
> 155
150 to 155
145 to 150
140 to 145
135 to 140
130 to 135
125 to 130
120 to 125
< 120

Hours of Sunshine
May

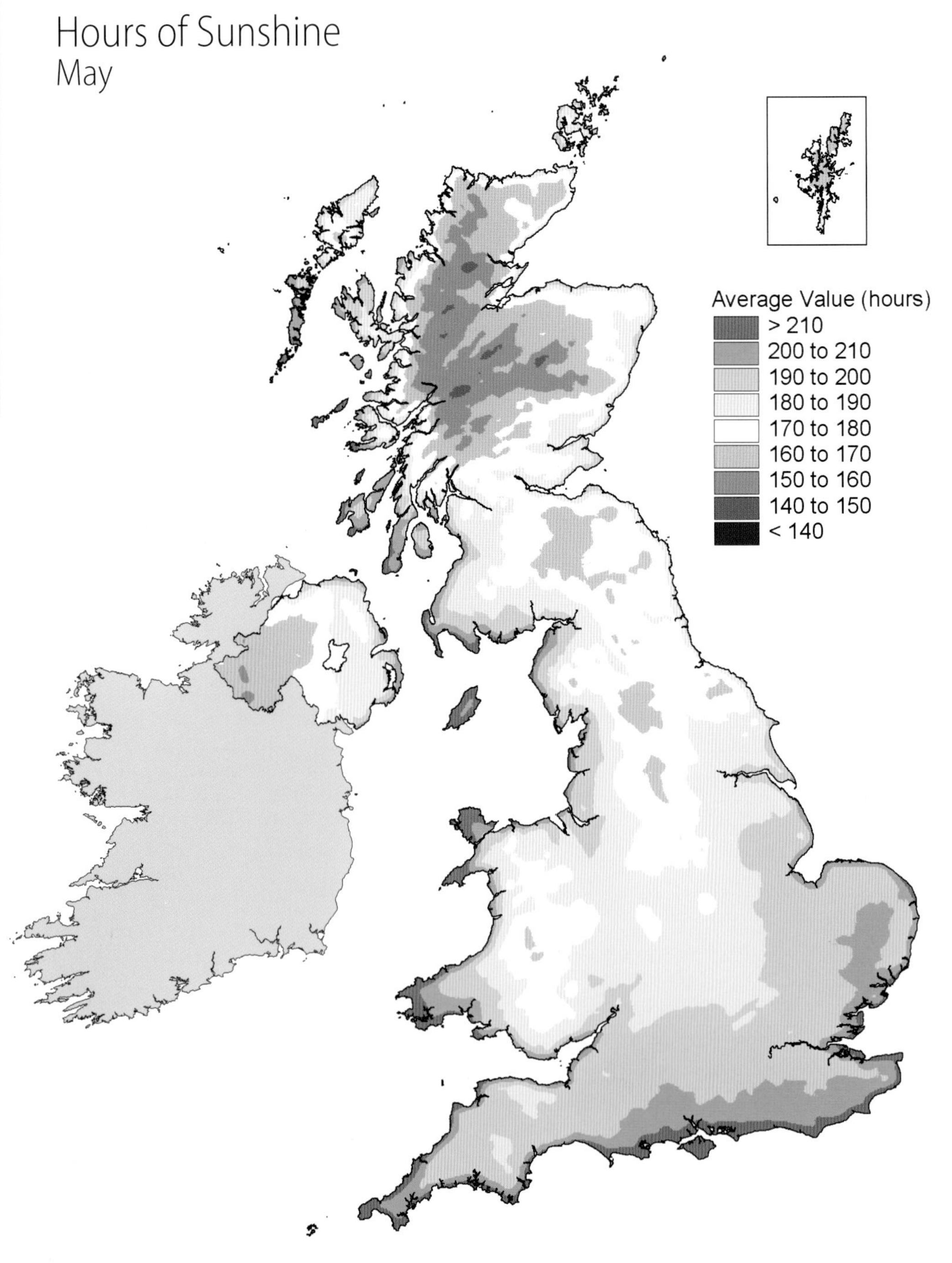
Average Value (hours)
> 210
200 to 210
190 to 200
180 to 190
170 to 180
160 to 170
150 to 160
140 to 150
< 140

Hours of Sunshine
June

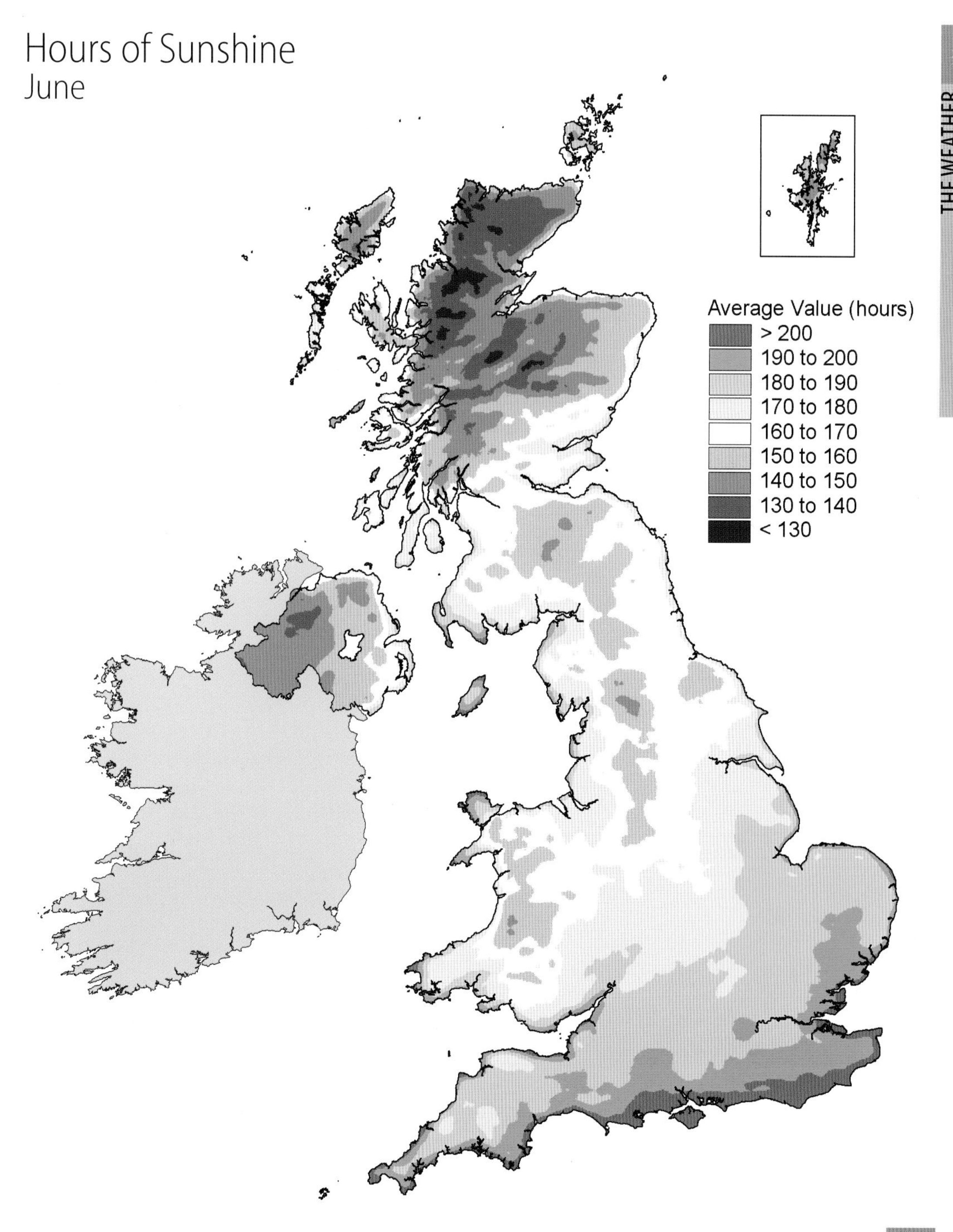
Average Value (hours)
> 200
190 to 200
180 to 190
170 to 180
160 to 170
150 to 160
140 to 150
130 to 140
< 130

Hours of Sunshine
July

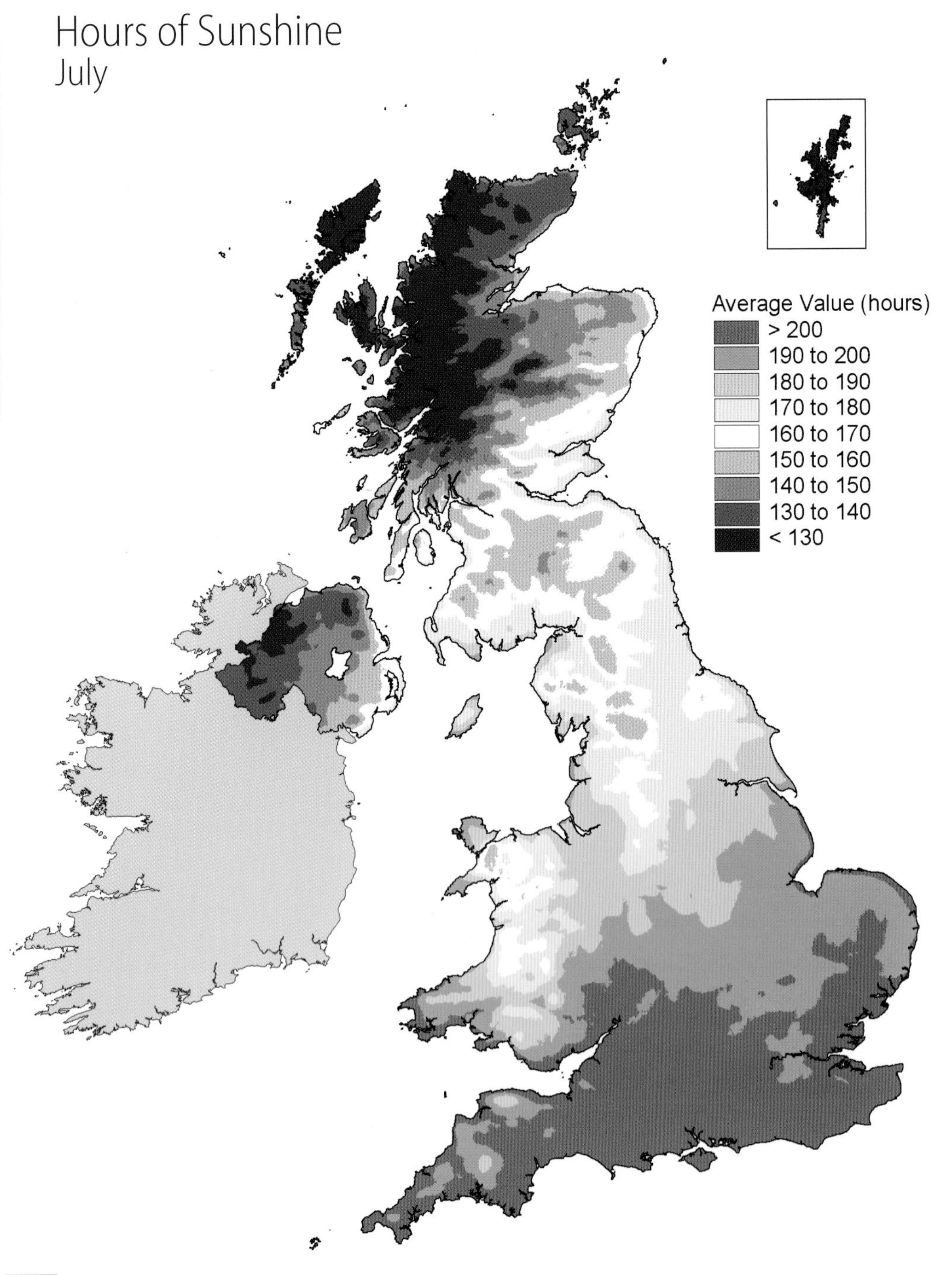
Average Value (hours)
> 200
190 to 200
180 to 190
170 to 180
160 to 170
150 to 160
140 to 150
130 to 140
< 130

Hours of Sunshine
August

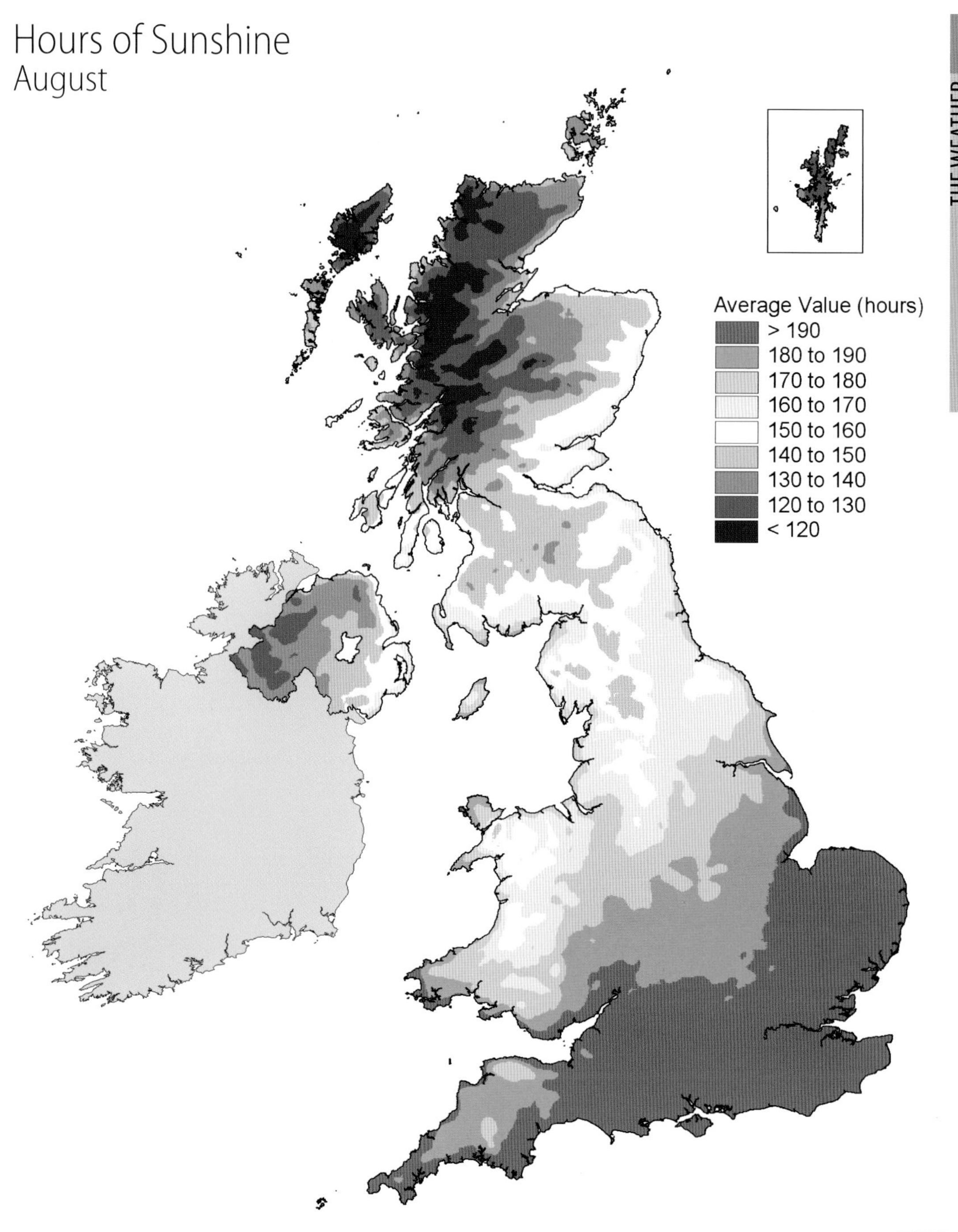

Average Value (hours)
> 190
180 to 190
170 to 180
160 to 170
150 to 160
140 to 150
130 to 140
120 to 130
< 120

Hours of Sunshine
September

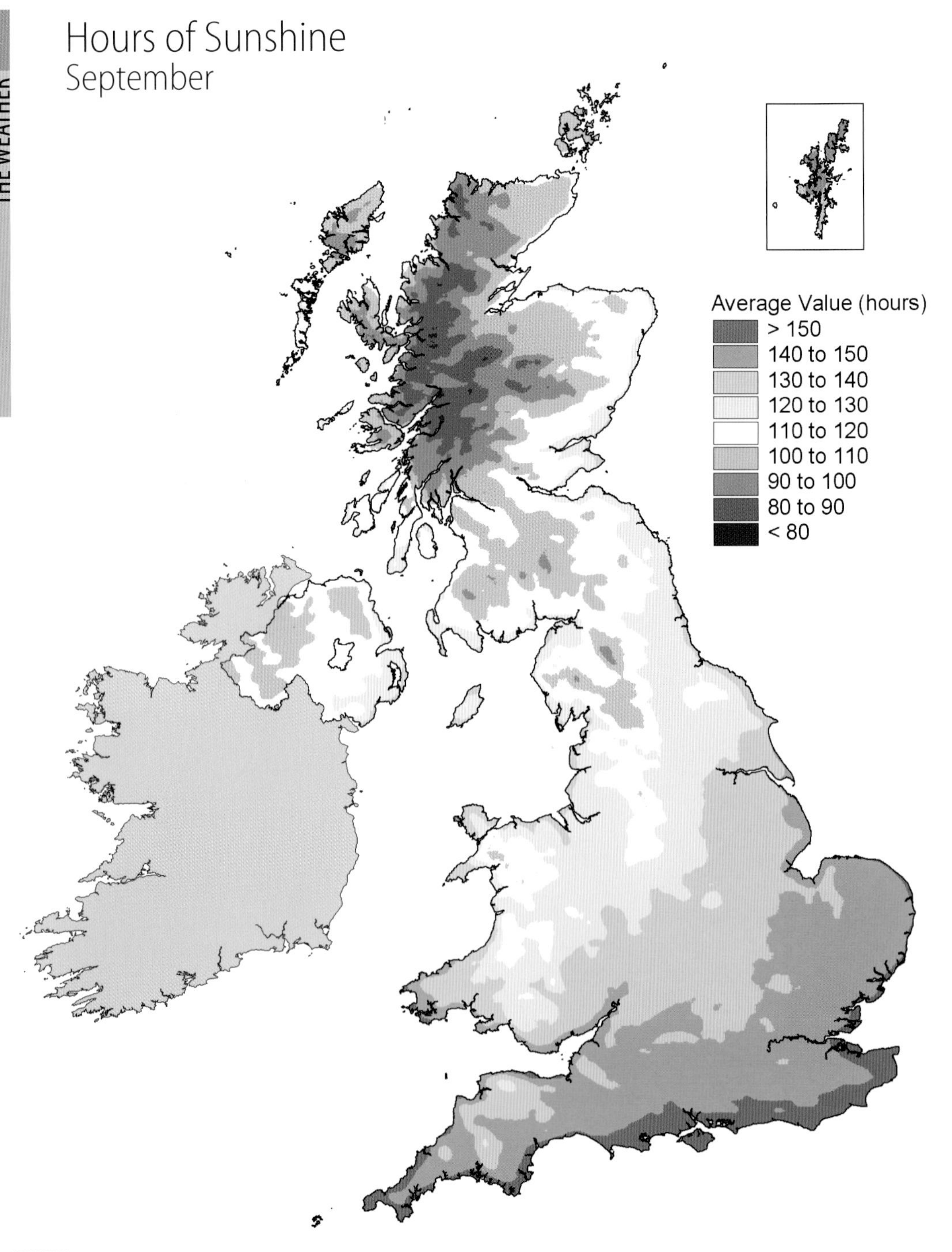
Average Value (hours)
> 150
140 to 150
130 to 140
120 to 130
110 to 120
100 to 110
90 to 100
80 to 90
< 80

Hours of Sunshine
October

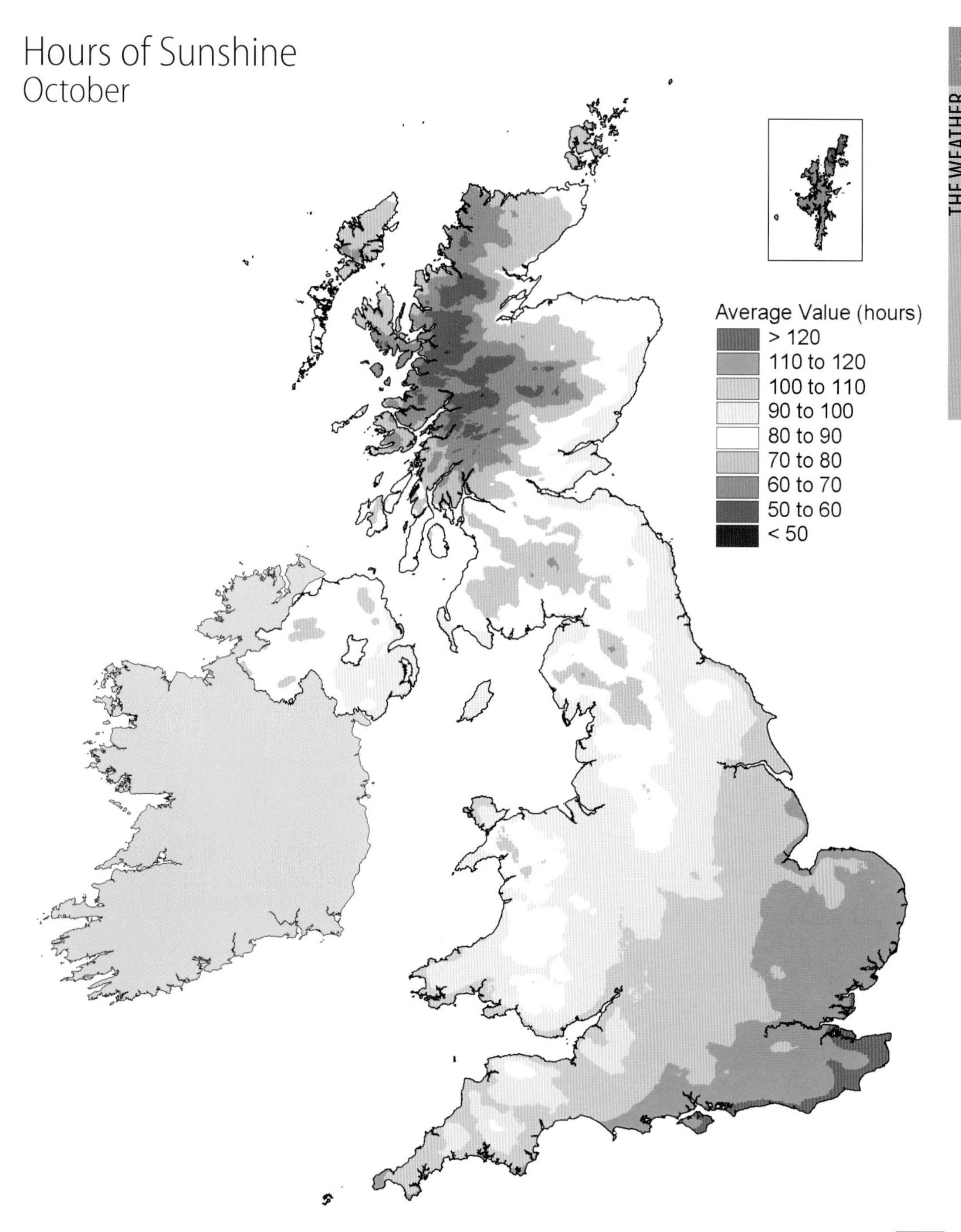

Average Value (hours)
> 120
110 to 120
100 to 110
90 to 100
80 to 90
70 to 80
60 to 70
50 to 60
< 50

Hours of Sunshine
November

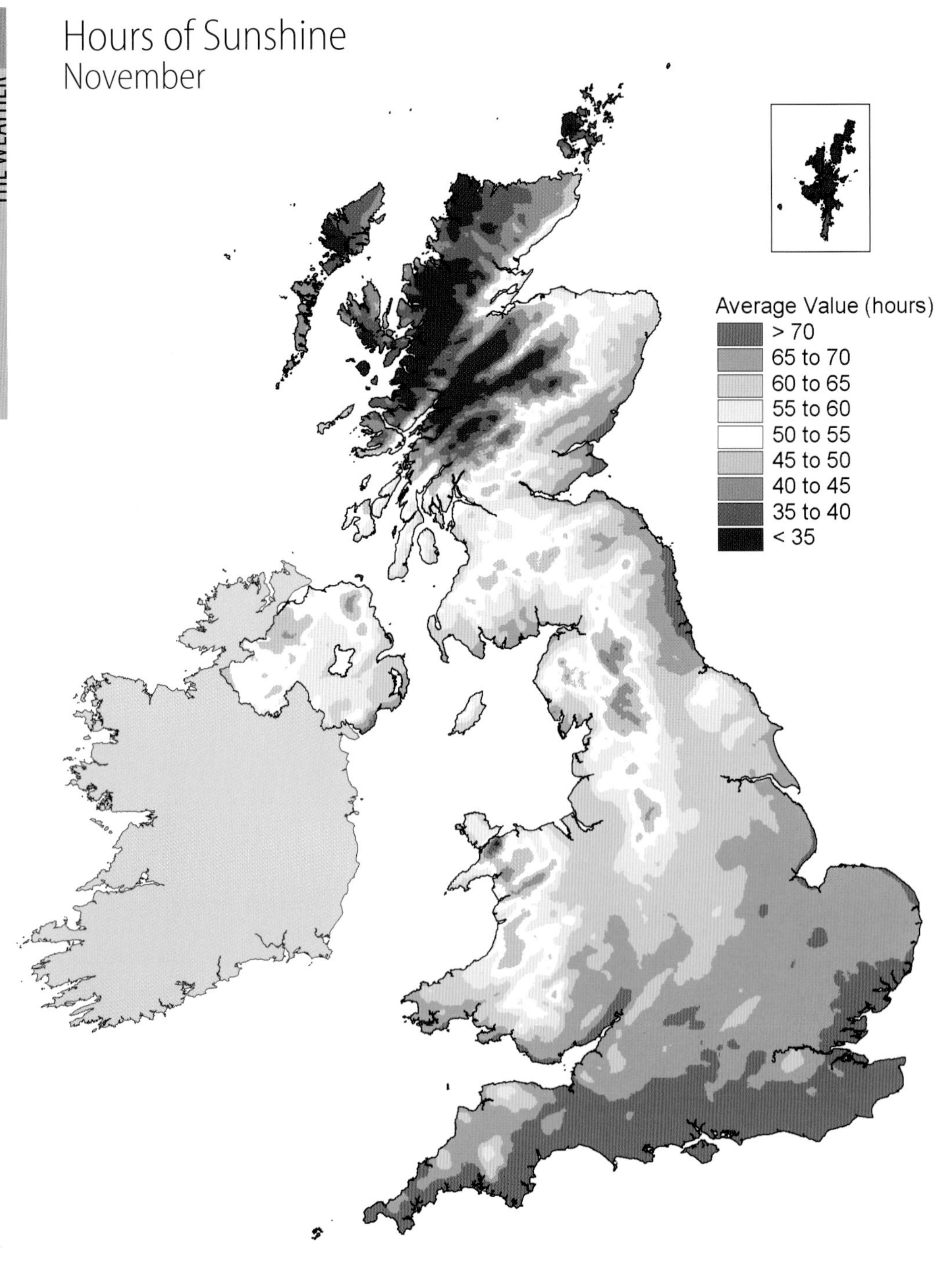
Average Value (hours)
> 70
65 to 70
60 to 65
55 to 60
50 to 55
45 to 50
40 to 45
35 to 40
< 35

Hours of Sunshine
December

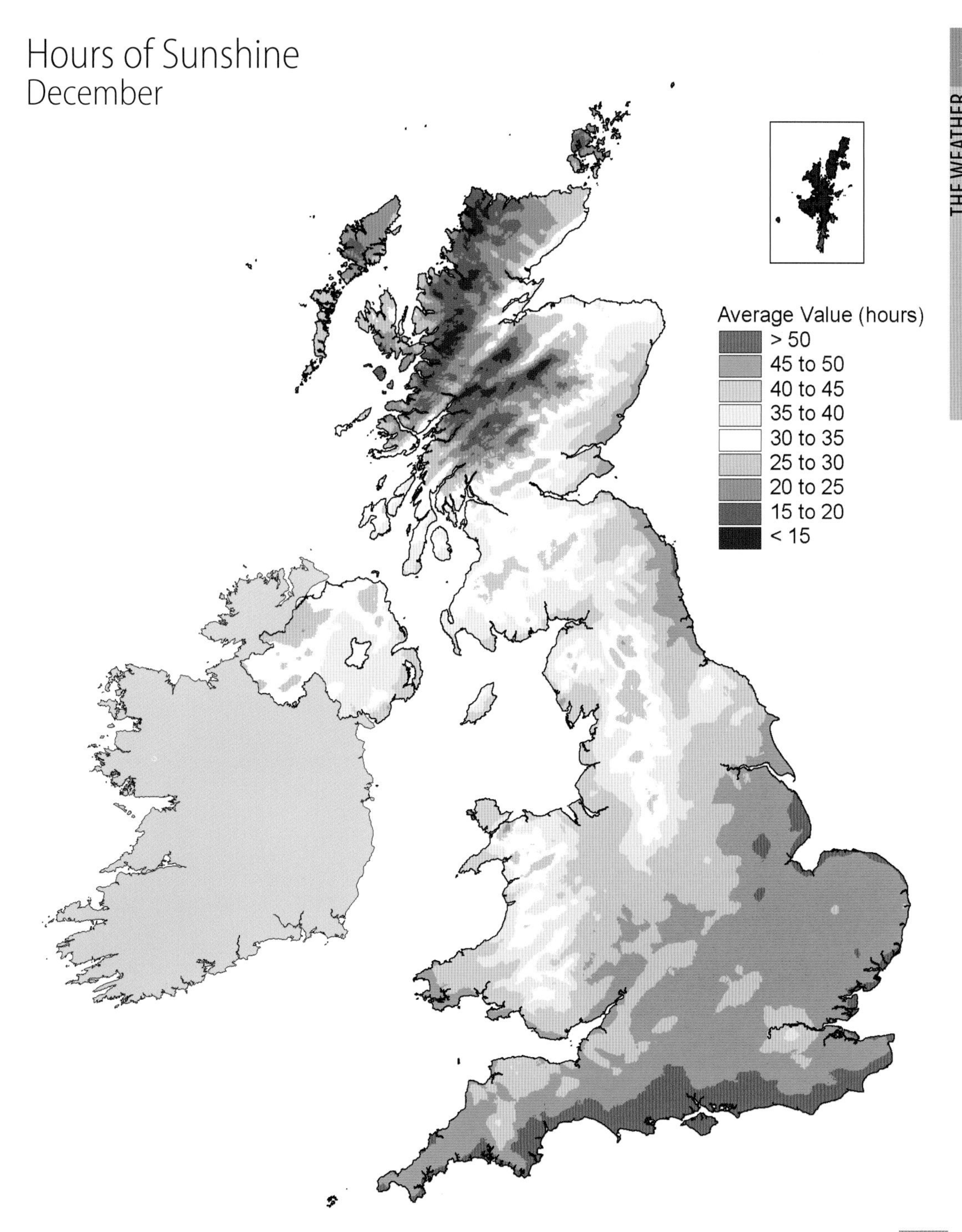

Average Value (hours)
> 50
45 to 50
40 to 45
35 to 40
30 to 35
25 to 30
20 to 25
15 to 20
< 15

Total Rainfall
Month-by-month

The wettest areas are over the mountains and moors of the north and west, which bear the brunt of the rain-bearing Atlantic weather systems. These are most active from October to January, giving the wet areas a distinct autumn or winter peak. In drier areas, rainfall is more evenly distributed throughout the year and in July and August can come in heavy showers. Wherever you are, April to June is normally the time when an umbrella is least needed.

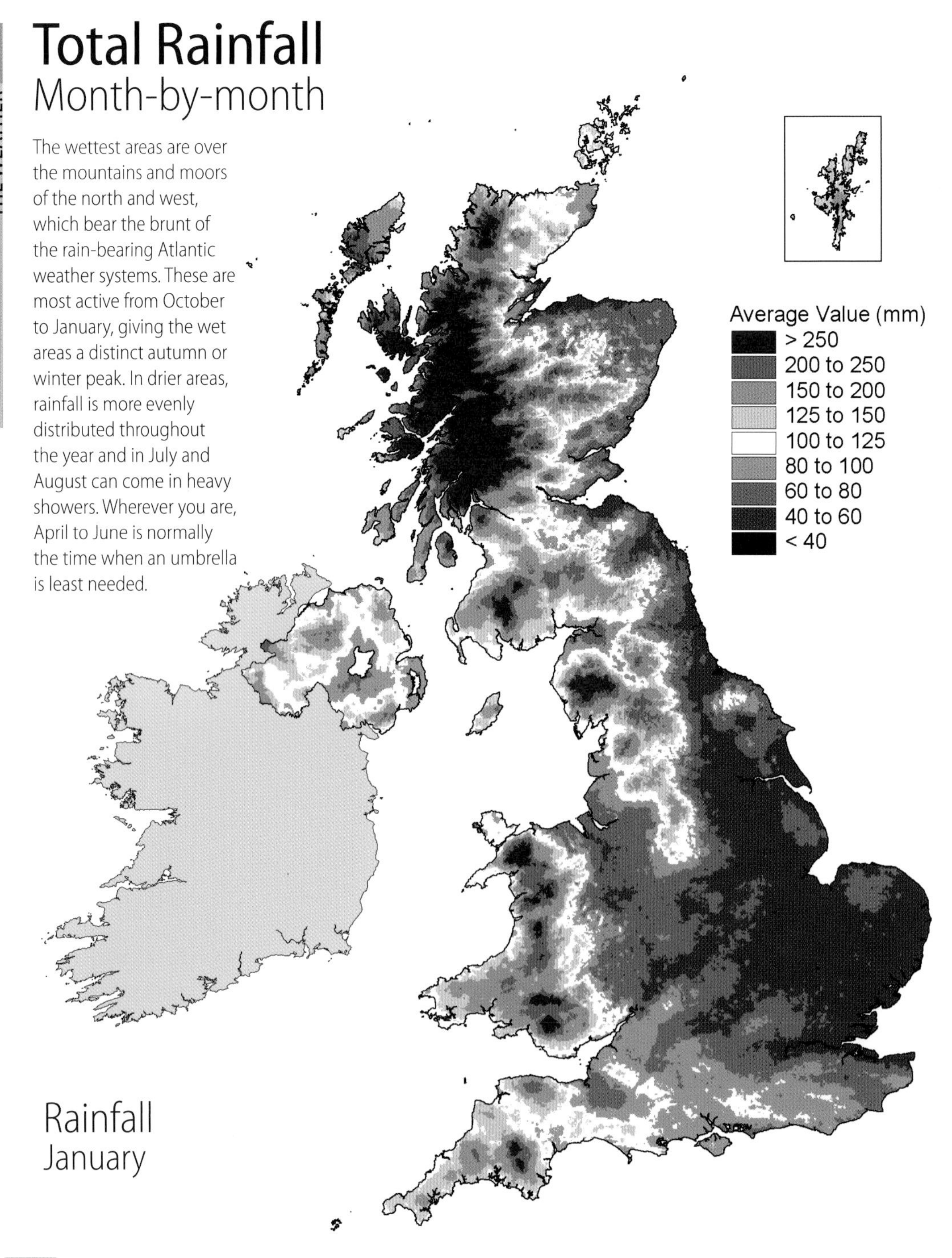
Average Value (mm)
> 250
200 to 250
150 to 200
125 to 150
100 to 125
80 to 100
60 to 80
40 to 60
< 40

Rainfall
January

Rainfall
February

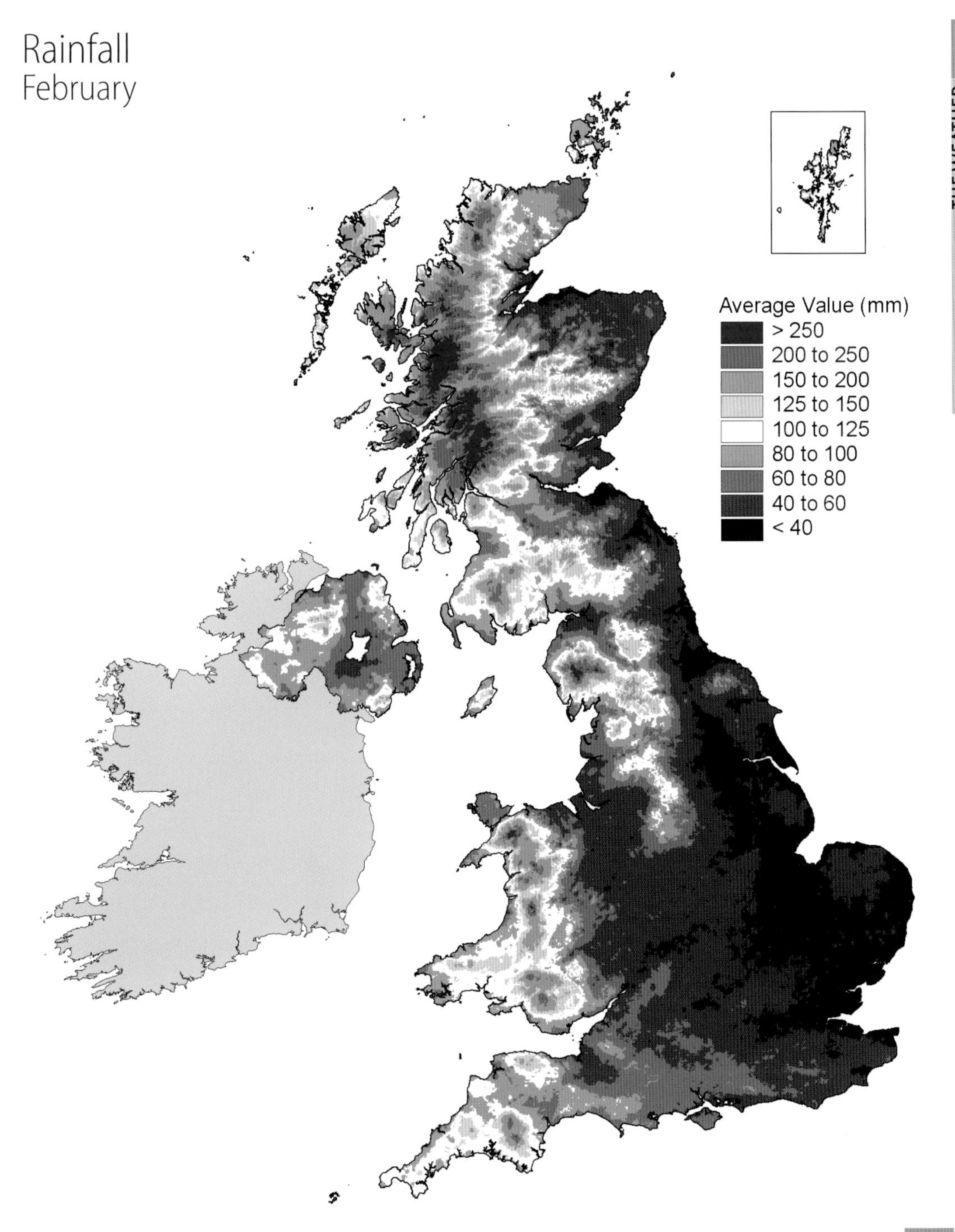
Average Value (mm)
> 250
200 to 250
150 to 200
125 to 150
100 to 125
80 to 100
60 to 80
40 to 60
< 40

Rainfall
March

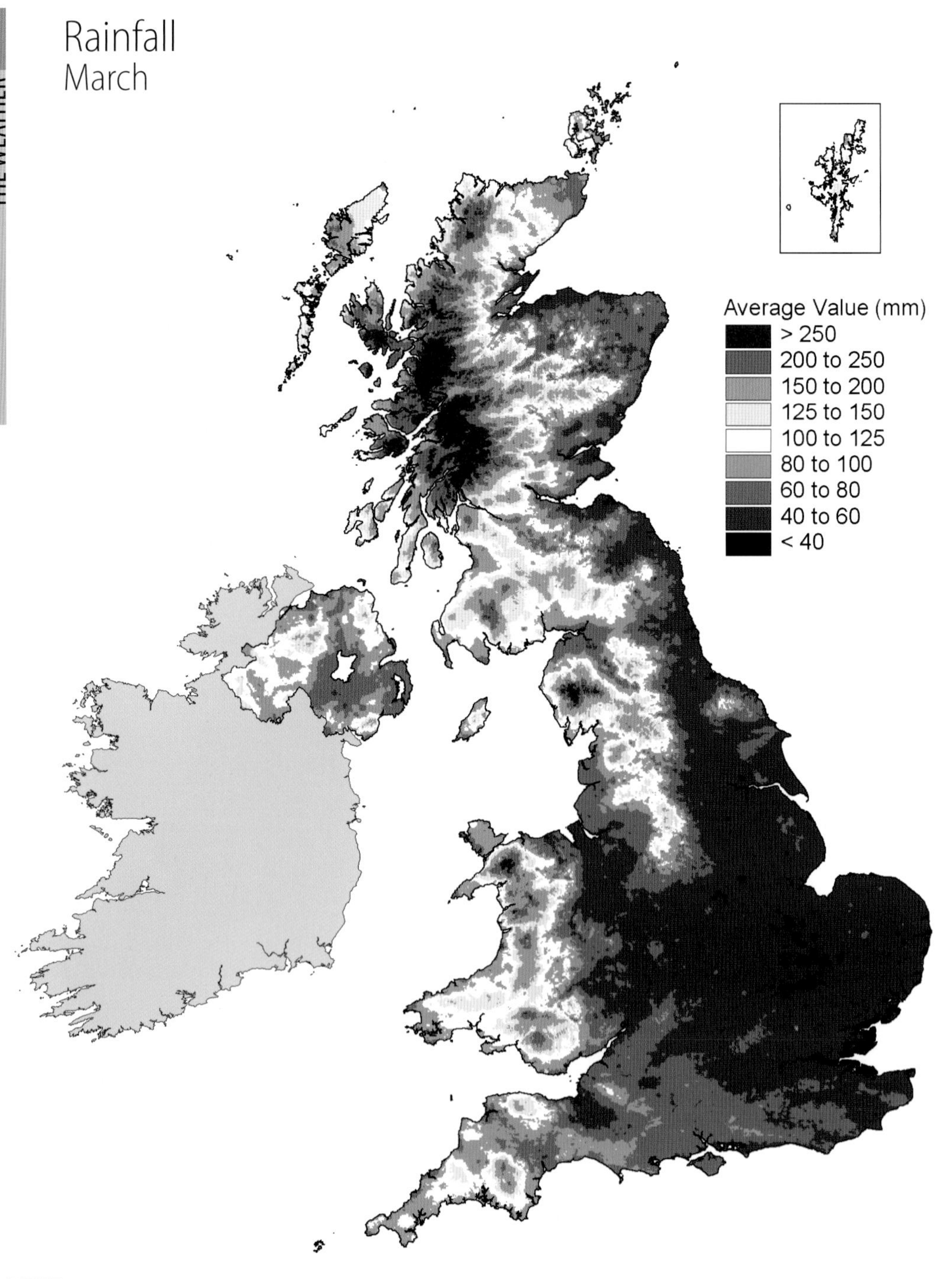
Average Value (mm)
> 250
200 to 250
150 to 200
125 to 150
100 to 125
80 to 100
60 to 80
40 to 60
< 40

Rainfall
April

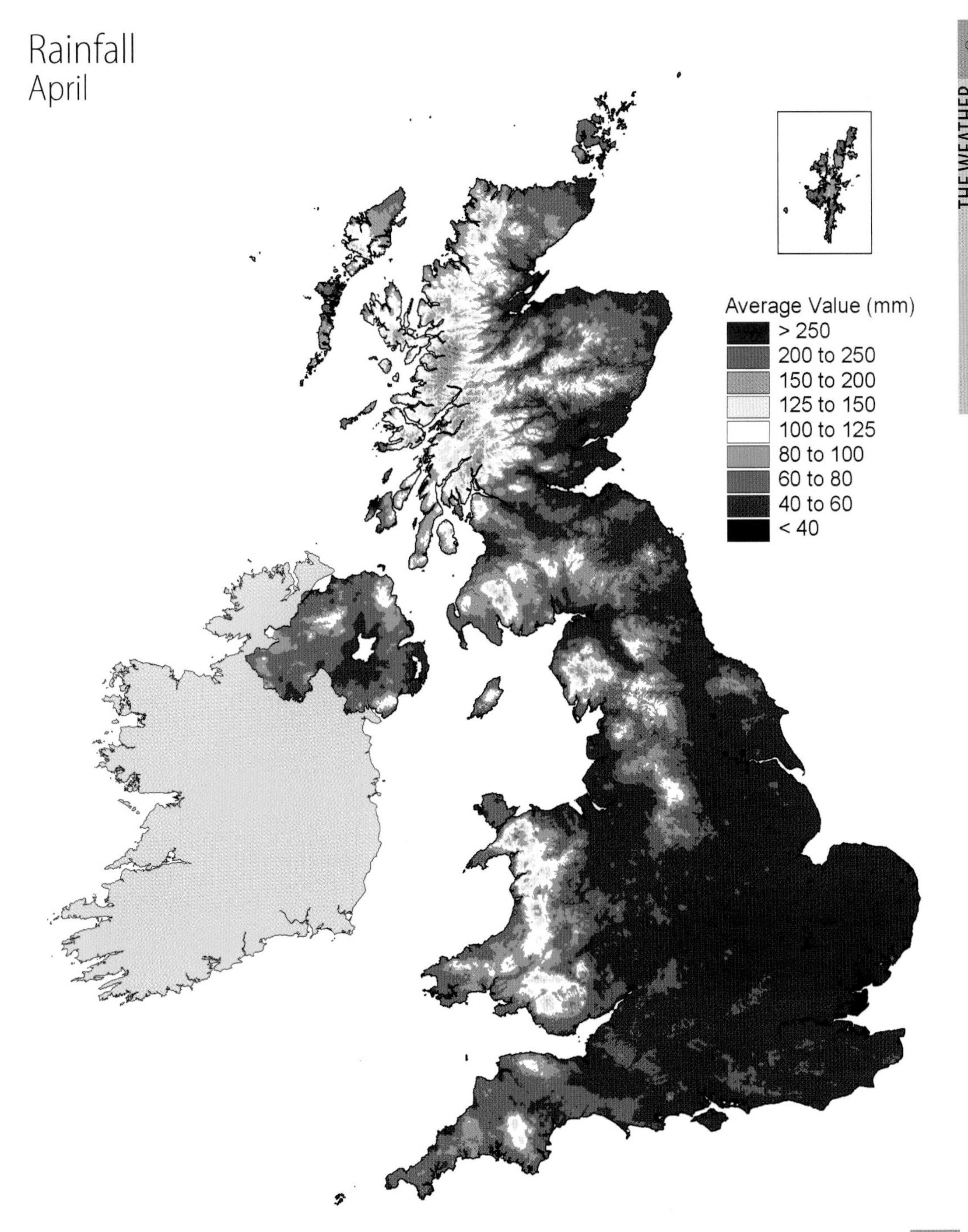

Average Value (mm)
> 250
200 to 250
150 to 200
125 to 150
100 to 125
80 to 100
60 to 80
40 to 60
< 40

Rainfall
May

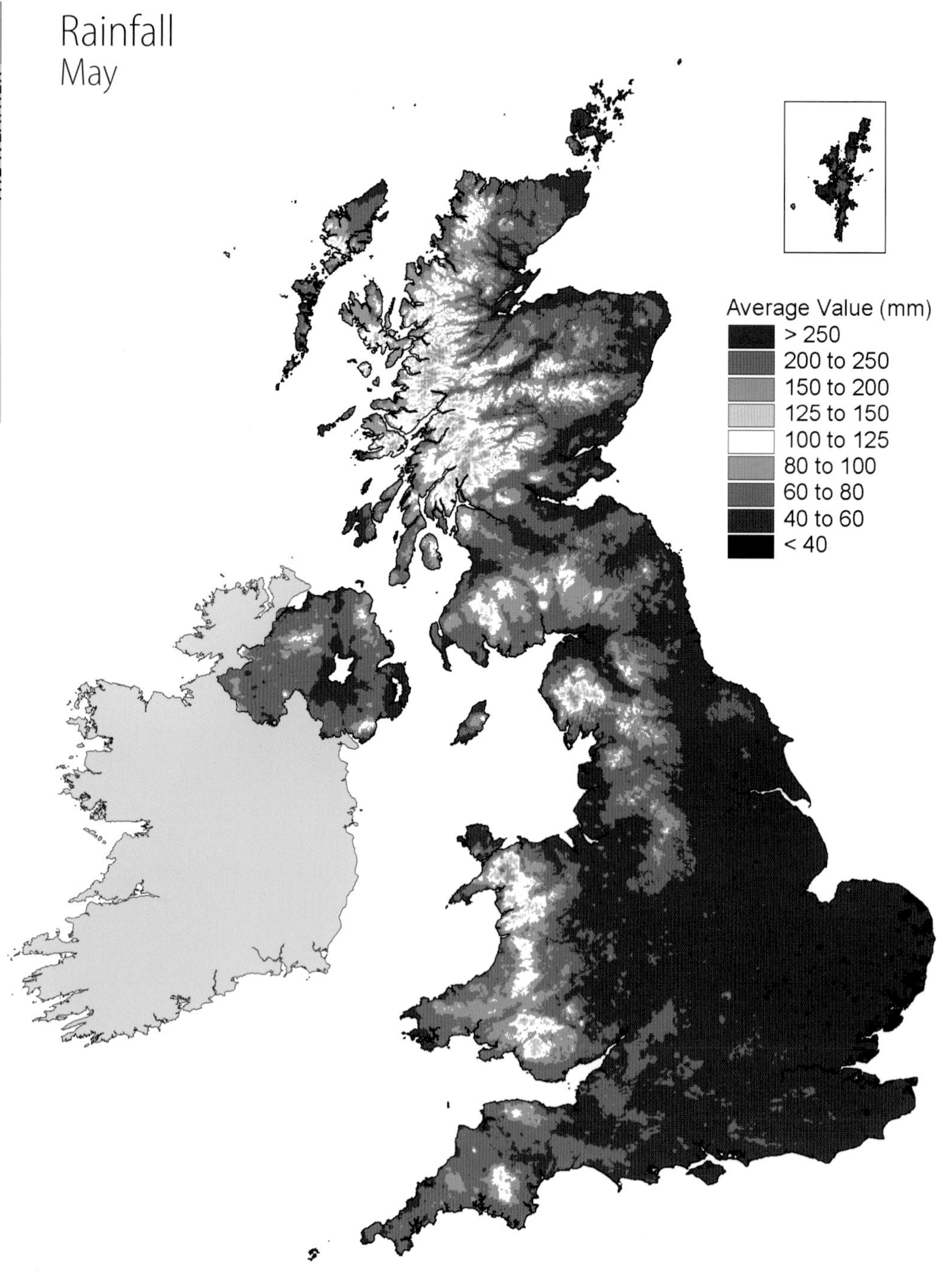
Average Value (mm)
> 250
200 to 250
150 to 200
125 to 150
100 to 125
80 to 100
60 to 80
40 to 60
< 40

Rainfall
June

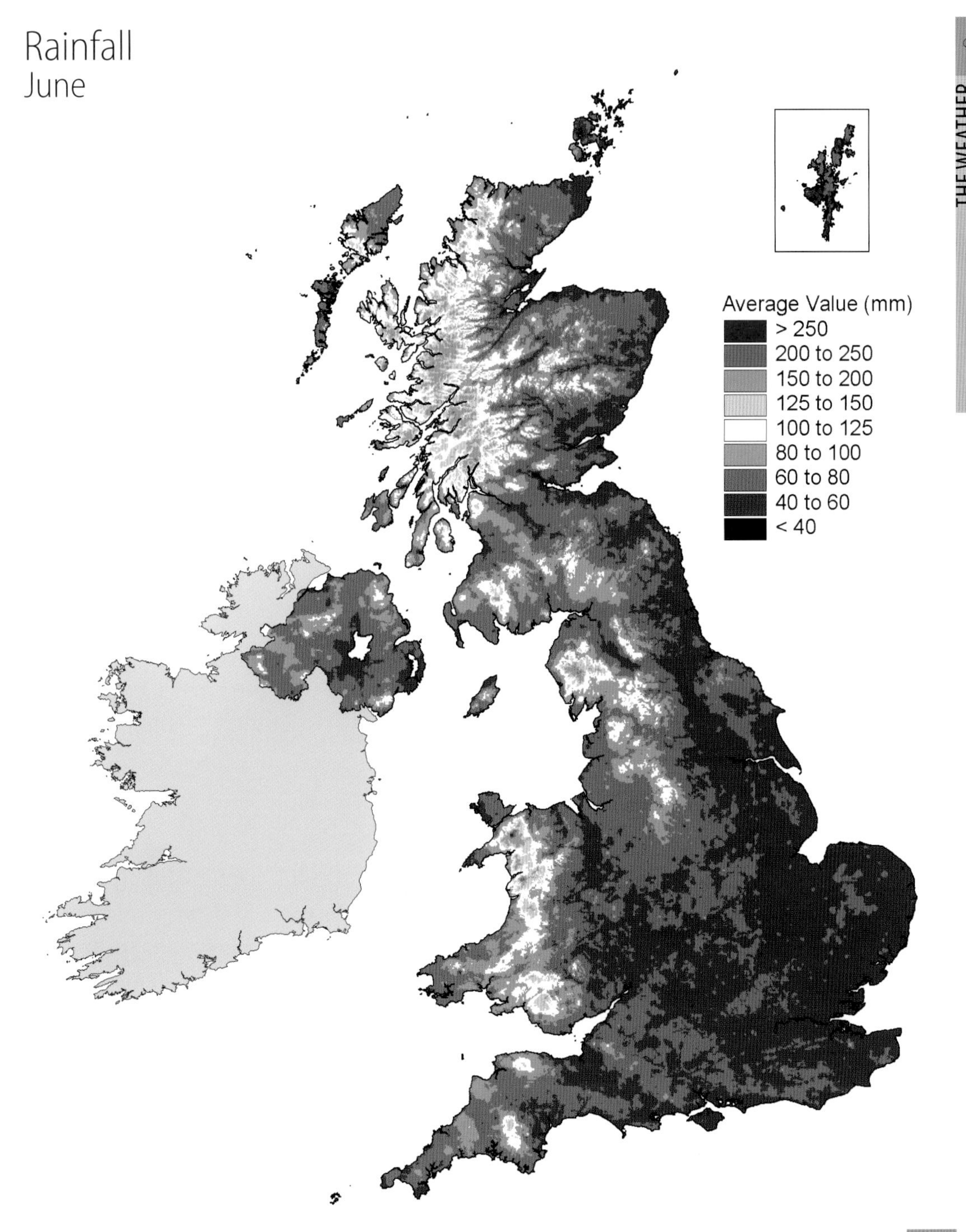
Average Value (mm)
> 250
200 to 250
150 to 200
125 to 150
100 to 125
80 to 100
60 to 80
40 to 60
< 40

Rainfall
July

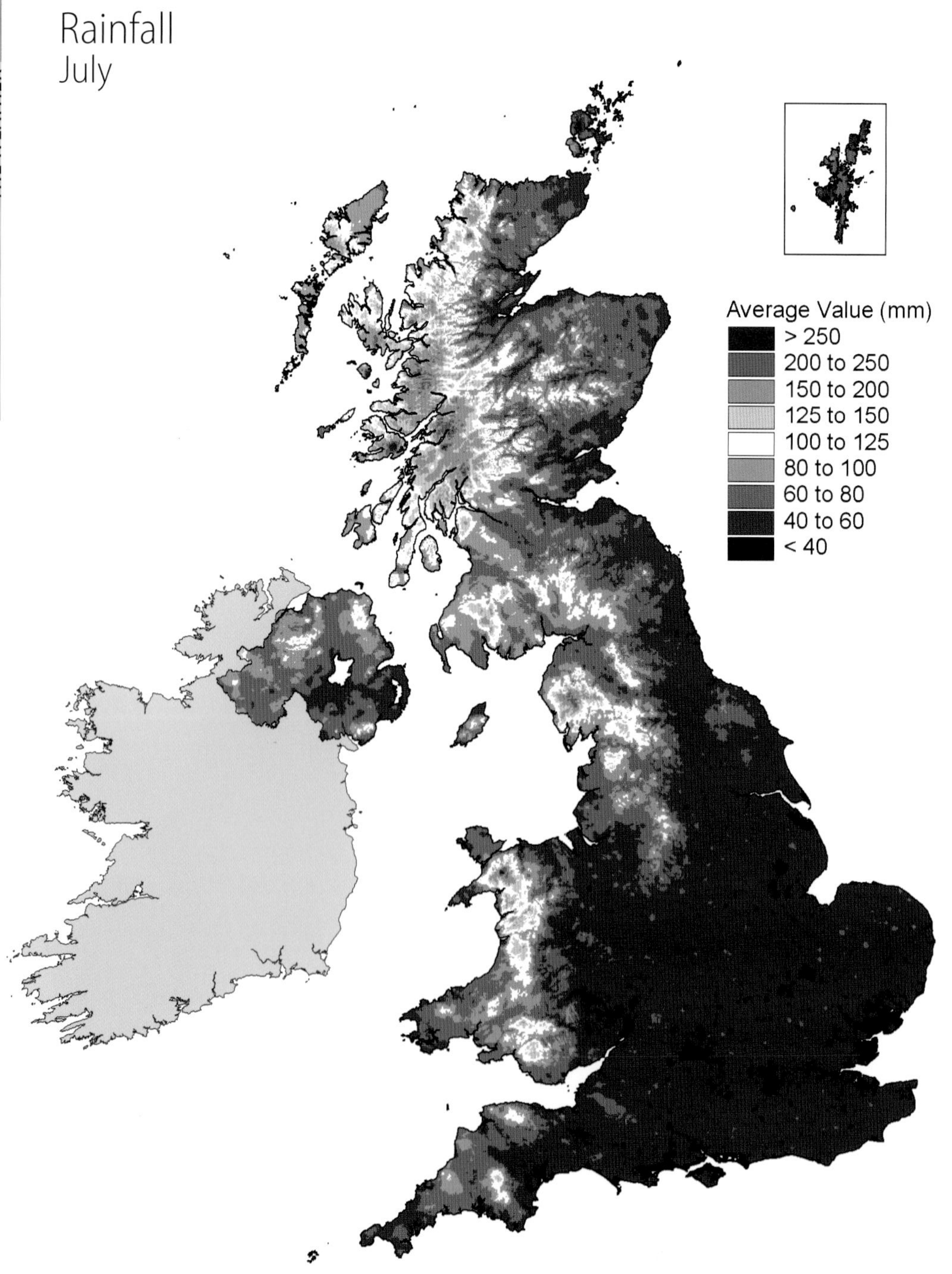
Average Value (mm)
> 250
200 to 250
150 to 200
125 to 150
100 to 125
80 to 100
60 to 80
40 to 60
< 40

Rainfall
August

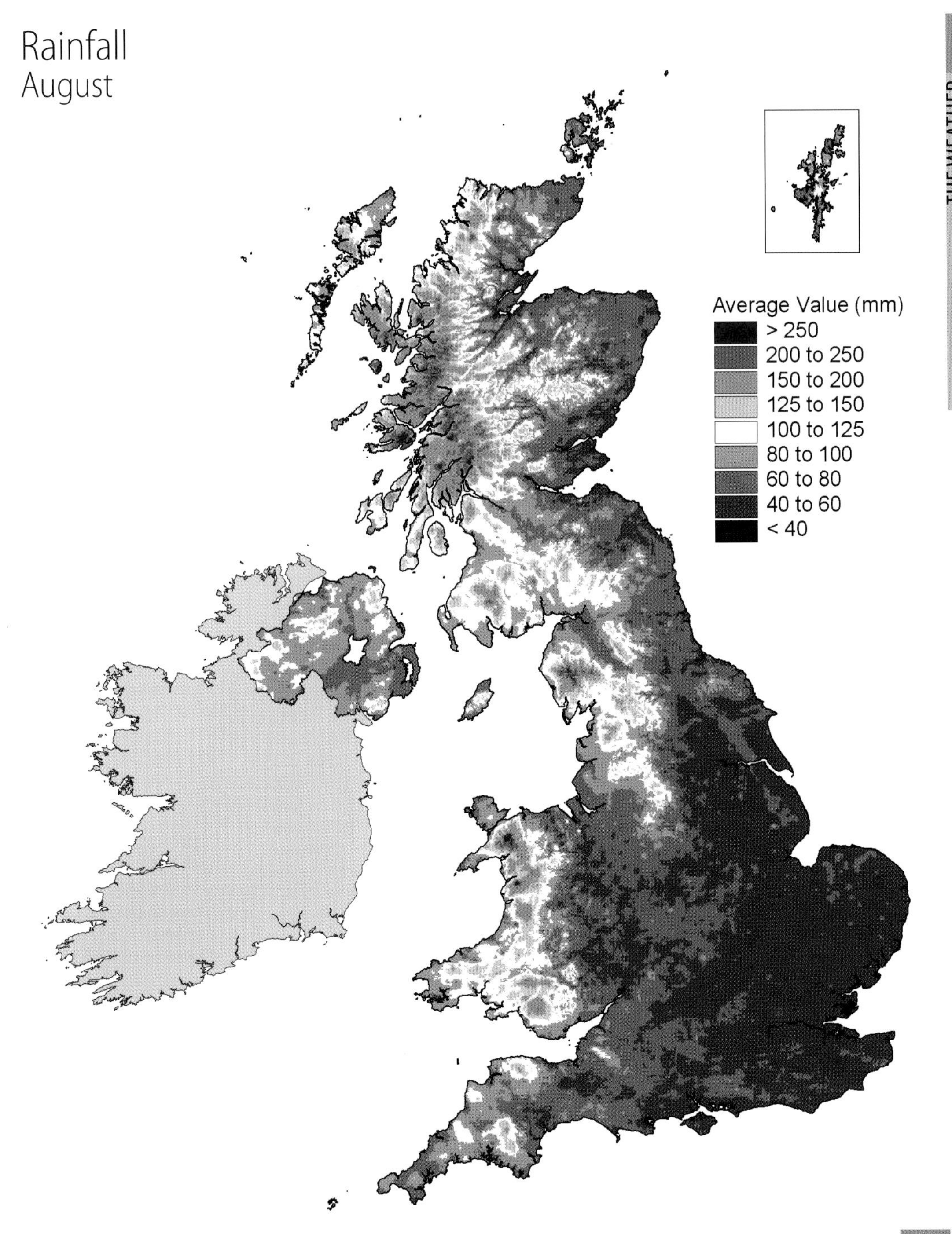
Average Value (mm)
> 250
200 to 250
150 to 200
125 to 150
100 to 125
80 to 100
60 to 80
40 to 60
< 40

Rainfall
September

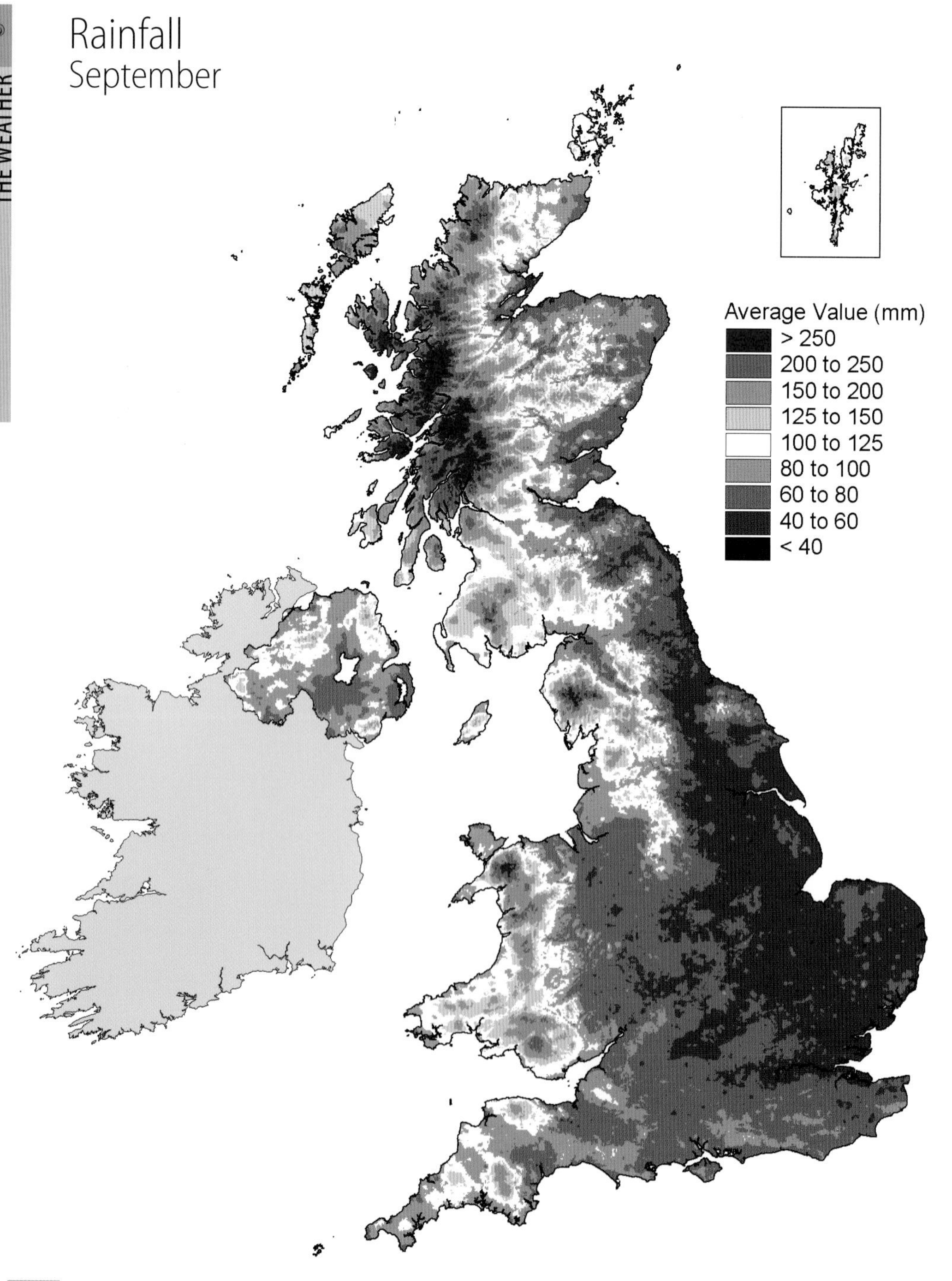
Average Value (mm)
> 250
200 to 250
150 to 200
125 to 150
100 to 125
80 to 100
60 to 80
40 to 60
< 40

Rainfall
October

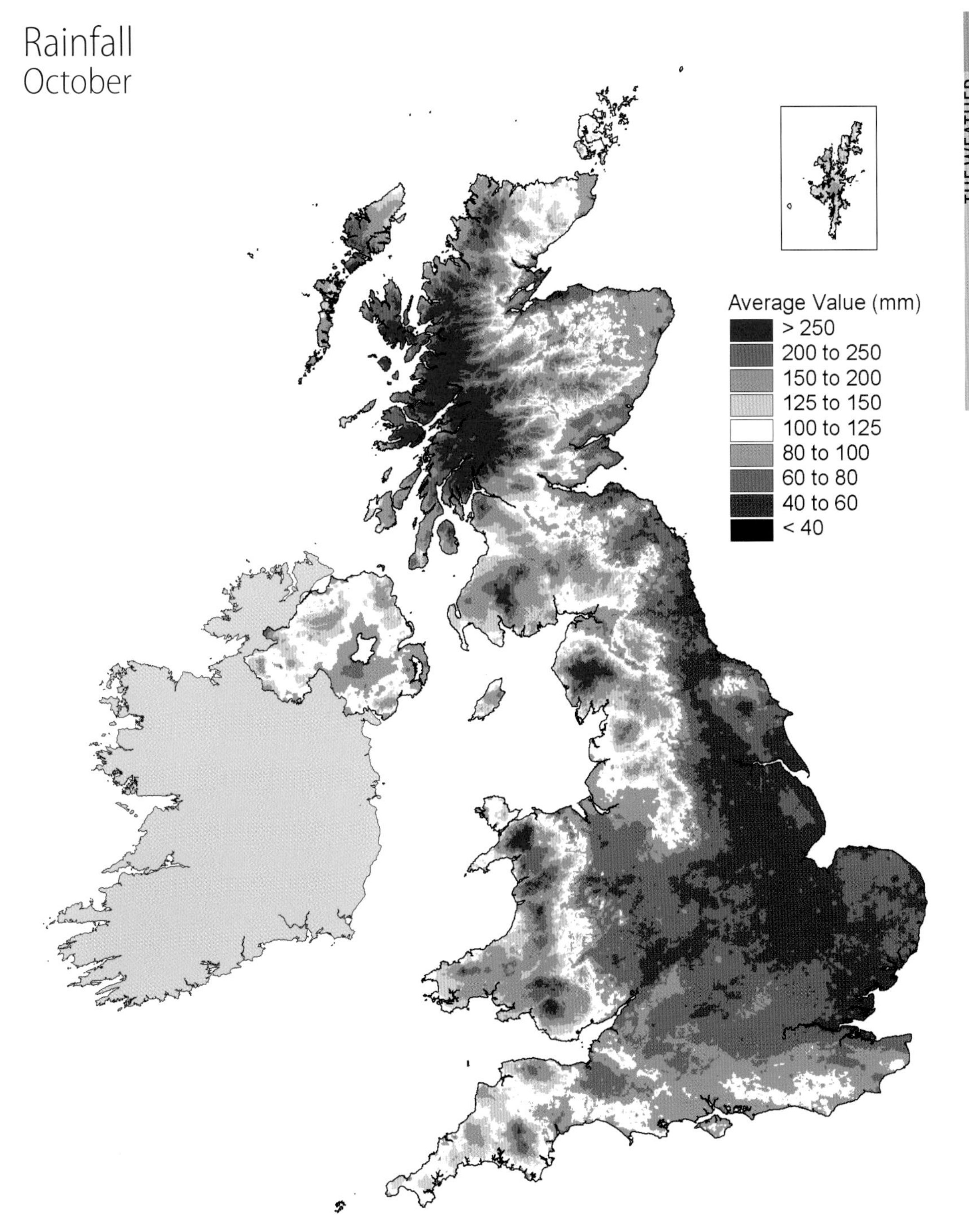

Average Value (mm)
> 250
200 to 250
150 to 200
125 to 150
100 to 125
80 to 100
60 to 80
40 to 60
< 40

Rainfall
November

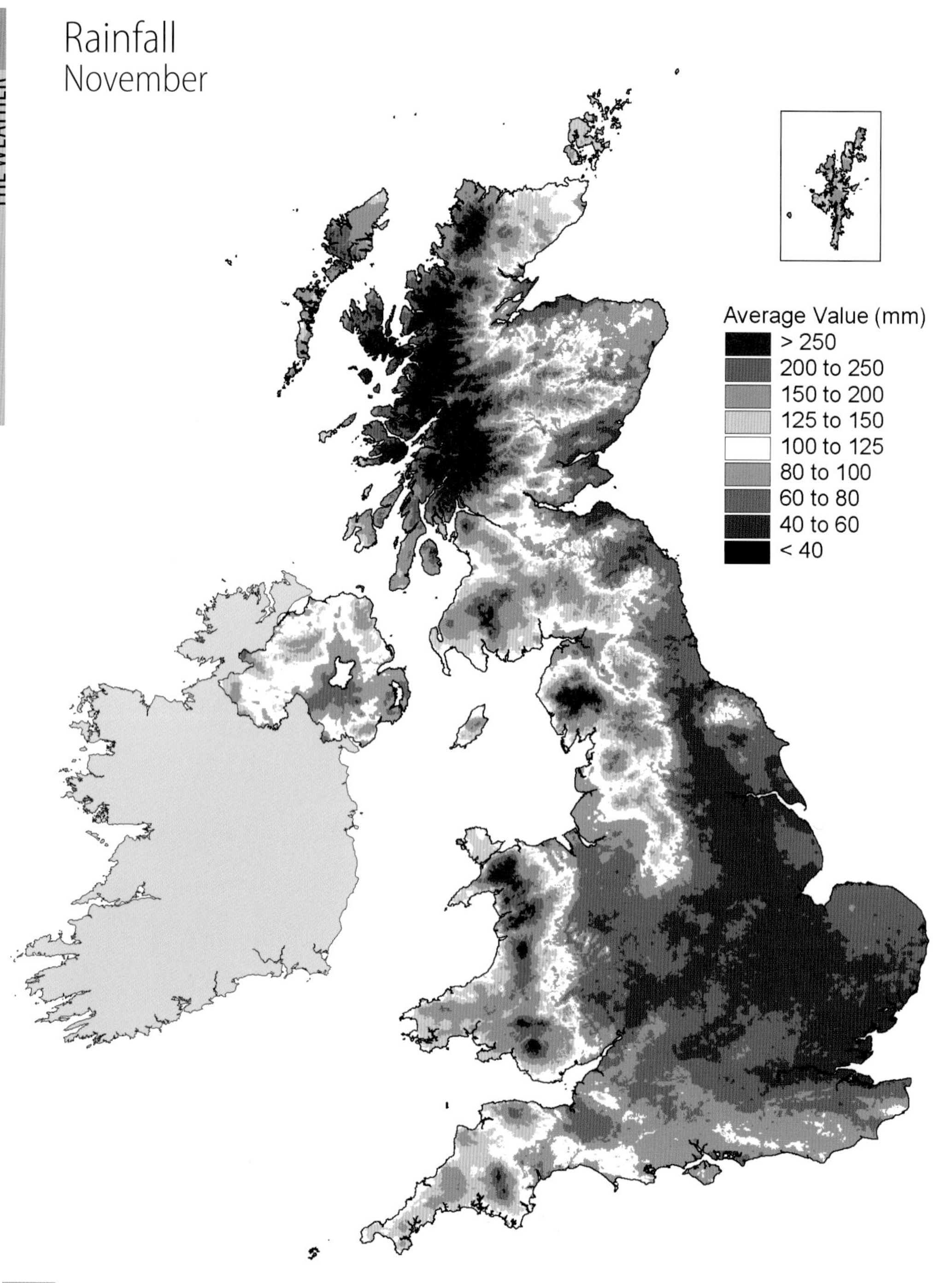

Average Value (mm)
> 250
200 to 250
150 to 200
125 to 150
100 to 125
80 to 100
60 to 80
40 to 60
< 40

Rainfall
December

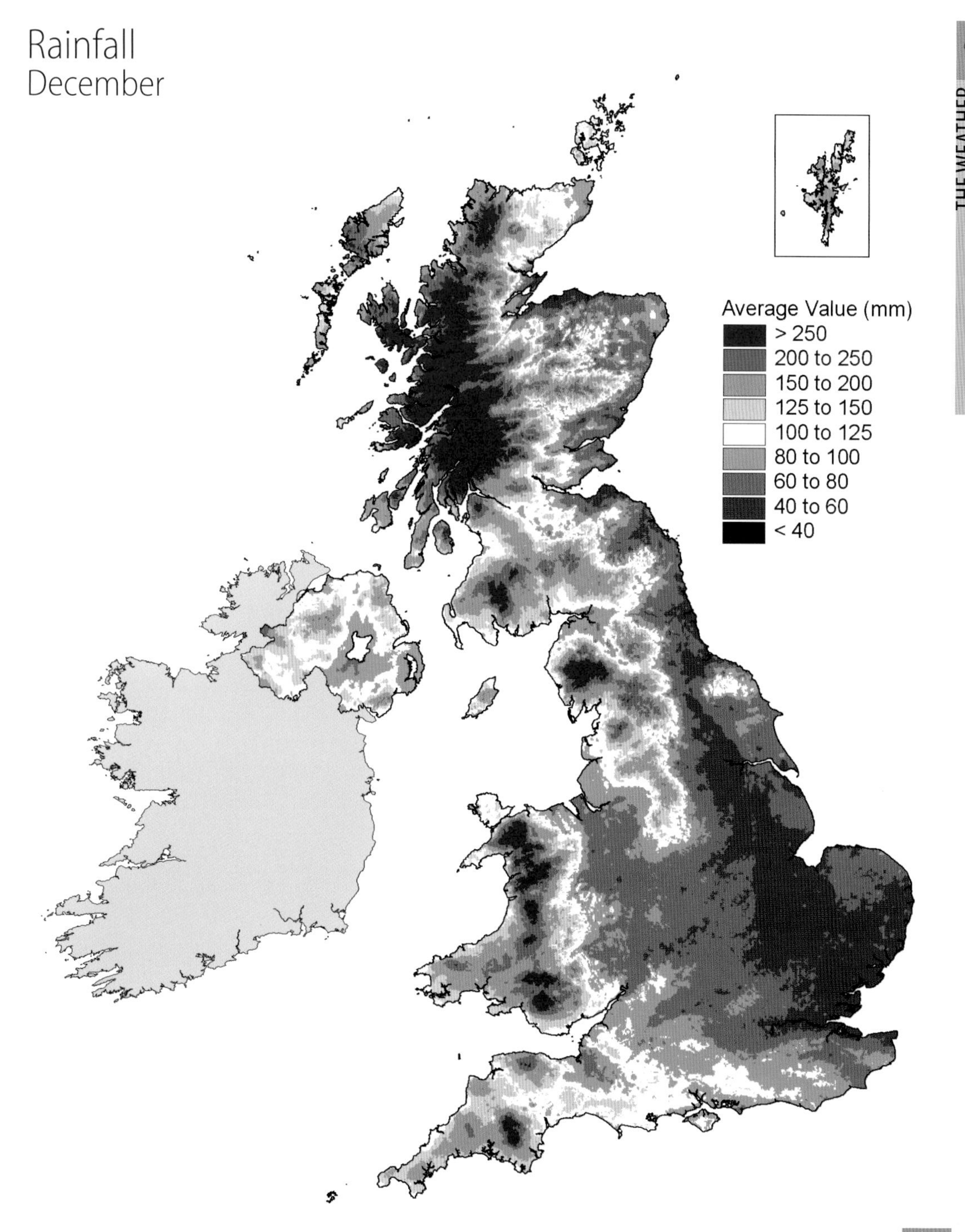

Average Value (mm)
> 250
200 to 250
150 to 200
125 to 150
100 to 125
80 to 100
60 to 80
40 to 60
< 40

Maximum Temperature
Month-by-month

In the winter months, daytime air temperatures tend to be highest in southern and western areas, close to the relatively warm waters of the Atlantic Ocean. January is the coldest month, when Cornwall averages 8°C – positively balmy compared to the near-freezing Scottish Highlands. In summer the south and east are favoured, especially if warm air arrives from the Continent. July is the warmest month, with mean values in the London area around 22°C, some 8°C above those in Shetland.

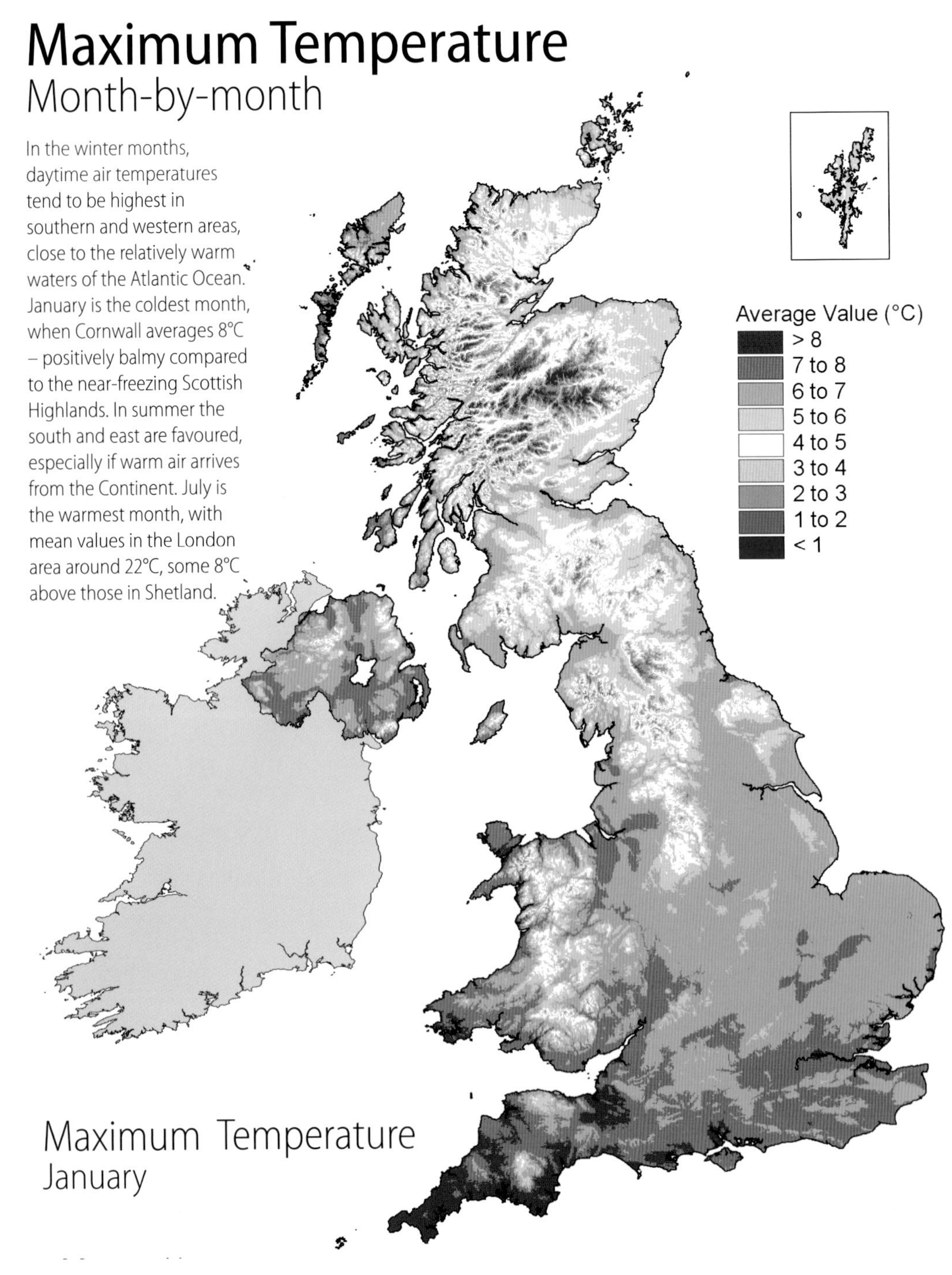

Average Value (°C)
> 8
7 to 8
6 to 7
5 to 6
4 to 5
3 to 4
2 to 3
1 to 2
< 1

Maximum Temperature
January

Maximum Temperature
February

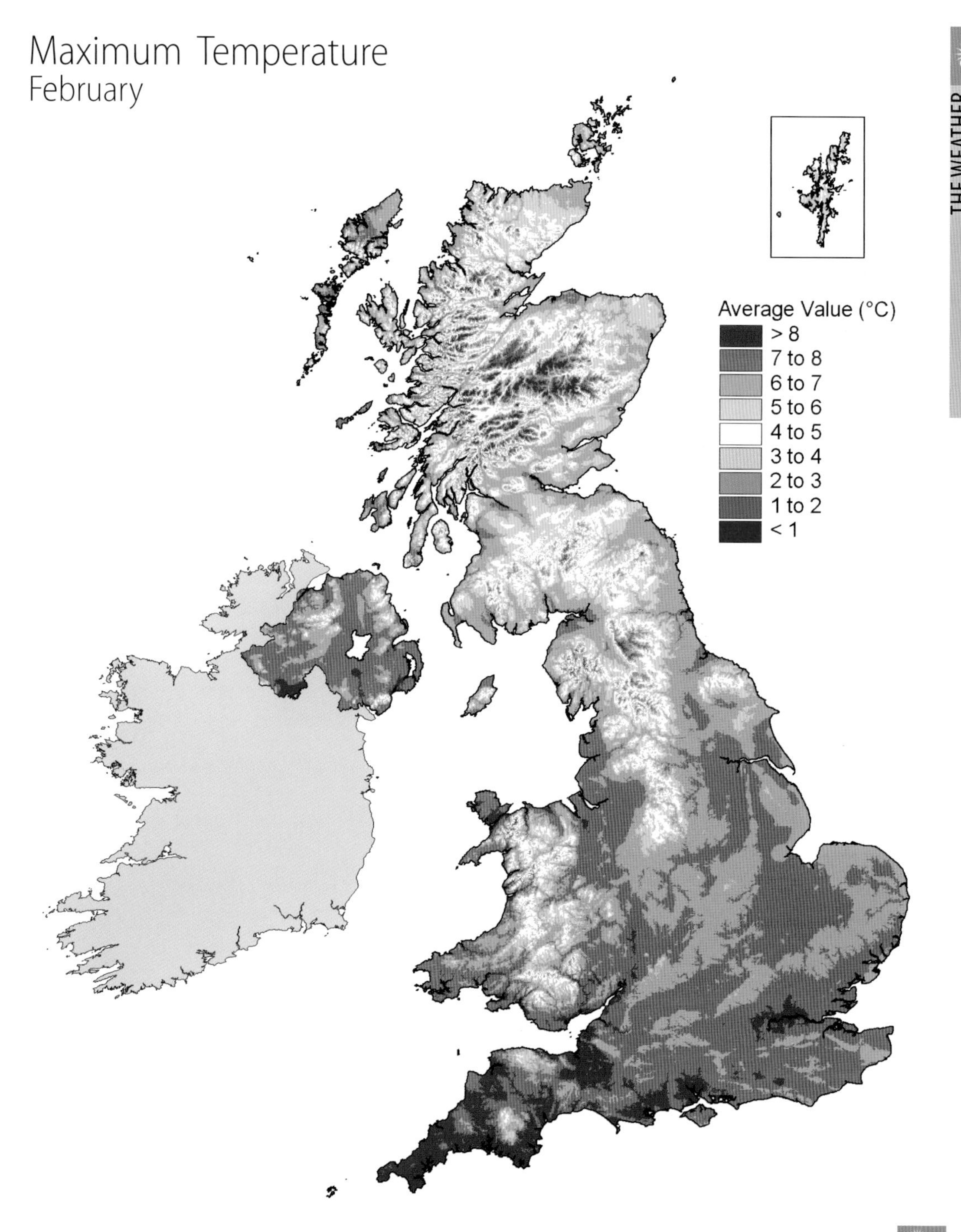
Average Value (°C)
> 8
7 to 8
6 to 7
5 to 6
4 to 5
3 to 4
2 to 3
1 to 2
< 1

Maximum Temperature
March

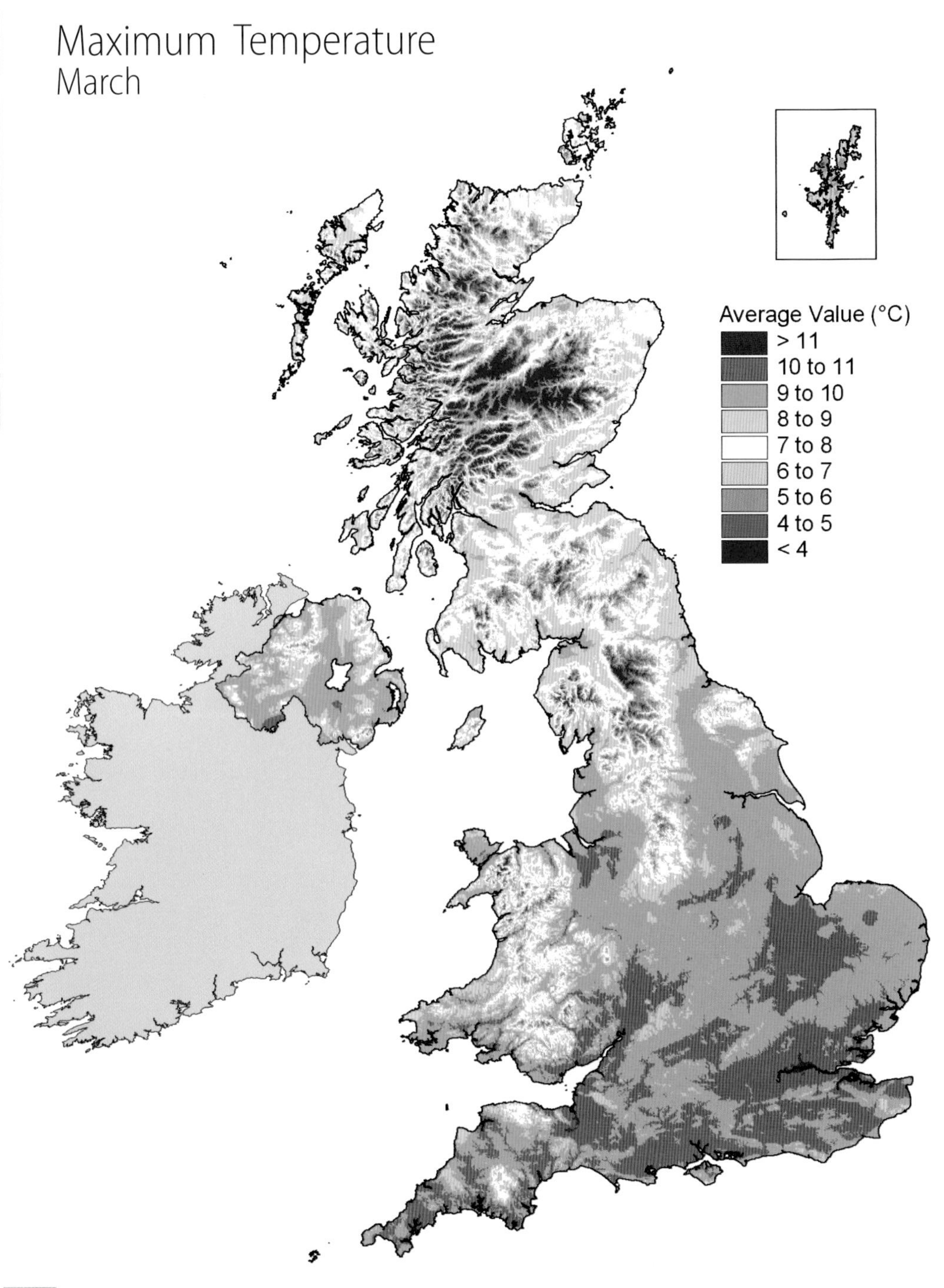
Average Value (°C)
> 11
10 to 11
9 to 10
8 to 9
7 to 8
6 to 7
5 to 6
4 to 5
< 4

Maximum Temperature
April

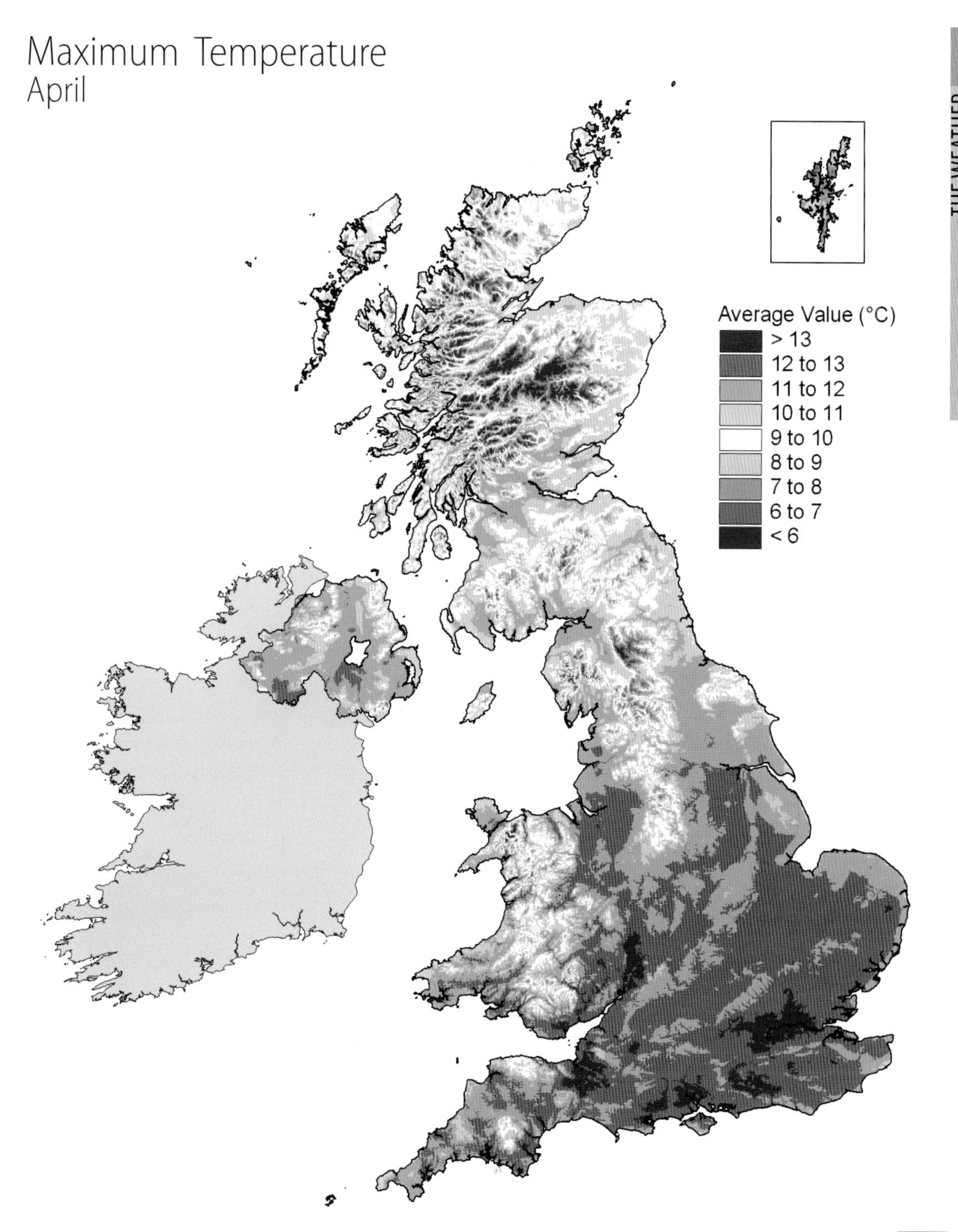

Average Value (°C)
> 13
12 to 13
11 to 12
10 to 11
9 to 10
8 to 9
7 to 8
6 to 7
< 6

Maximum Temperature
May

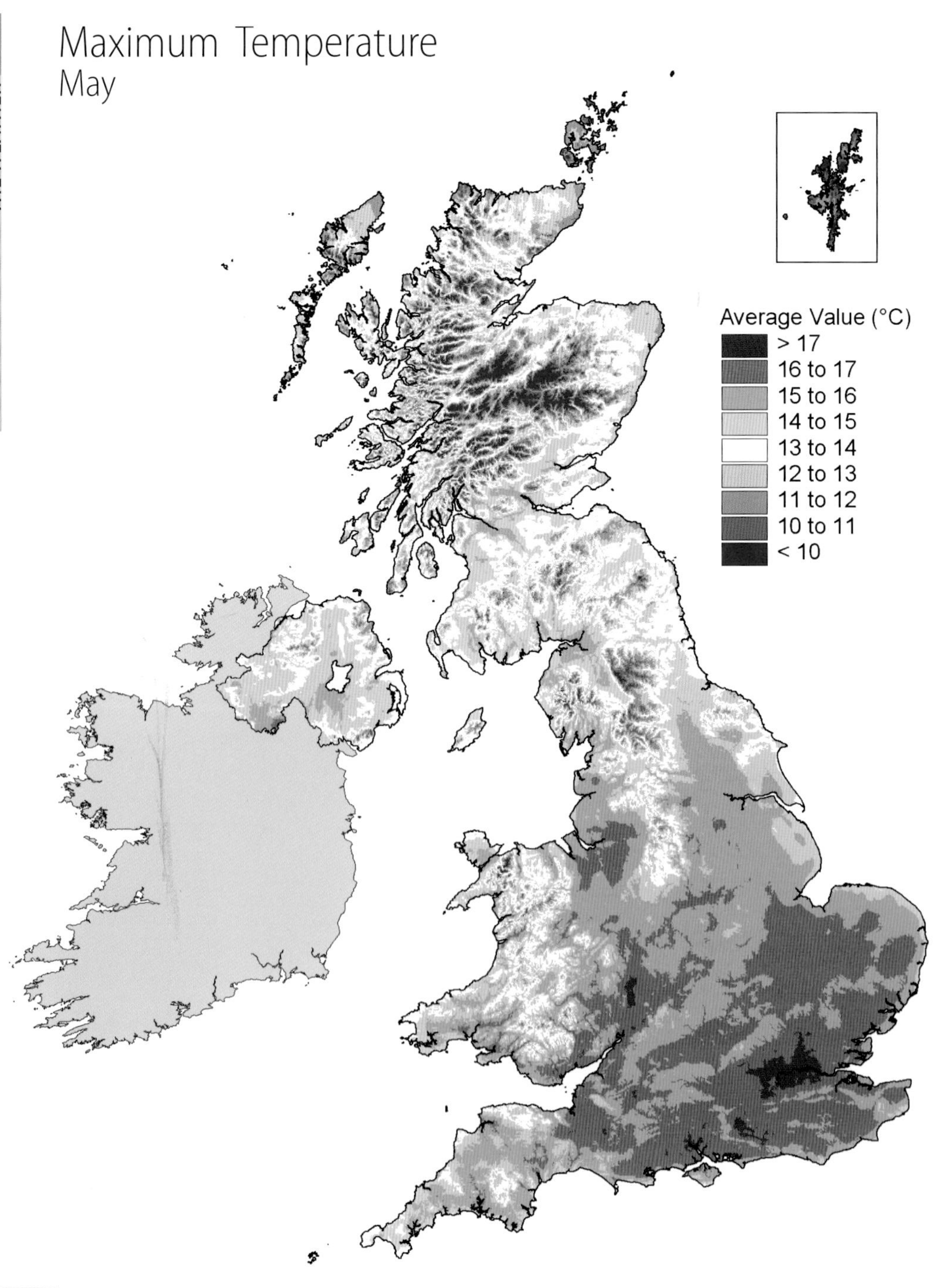
Average Value (°C)
> 17
16 to 17
15 to 16
14 to 15
13 to 14
12 to 13
11 to 12
10 to 11
< 10

Maximum Temperature
June

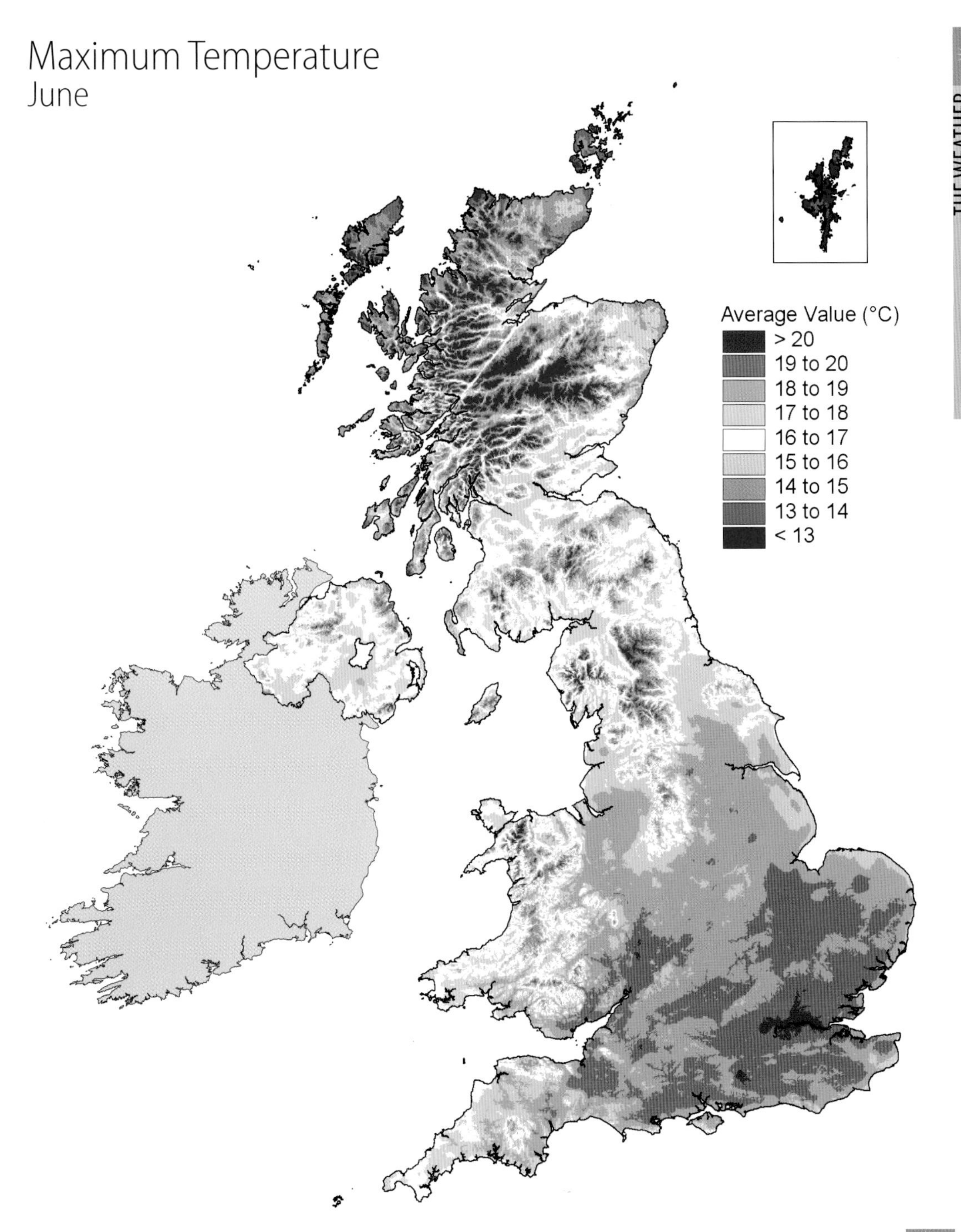
Average Value (°C)
> 20
19 to 20
18 to 19
17 to 18
16 to 17
15 to 16
14 to 15
13 to 14
< 13

Maximum Temperature
July

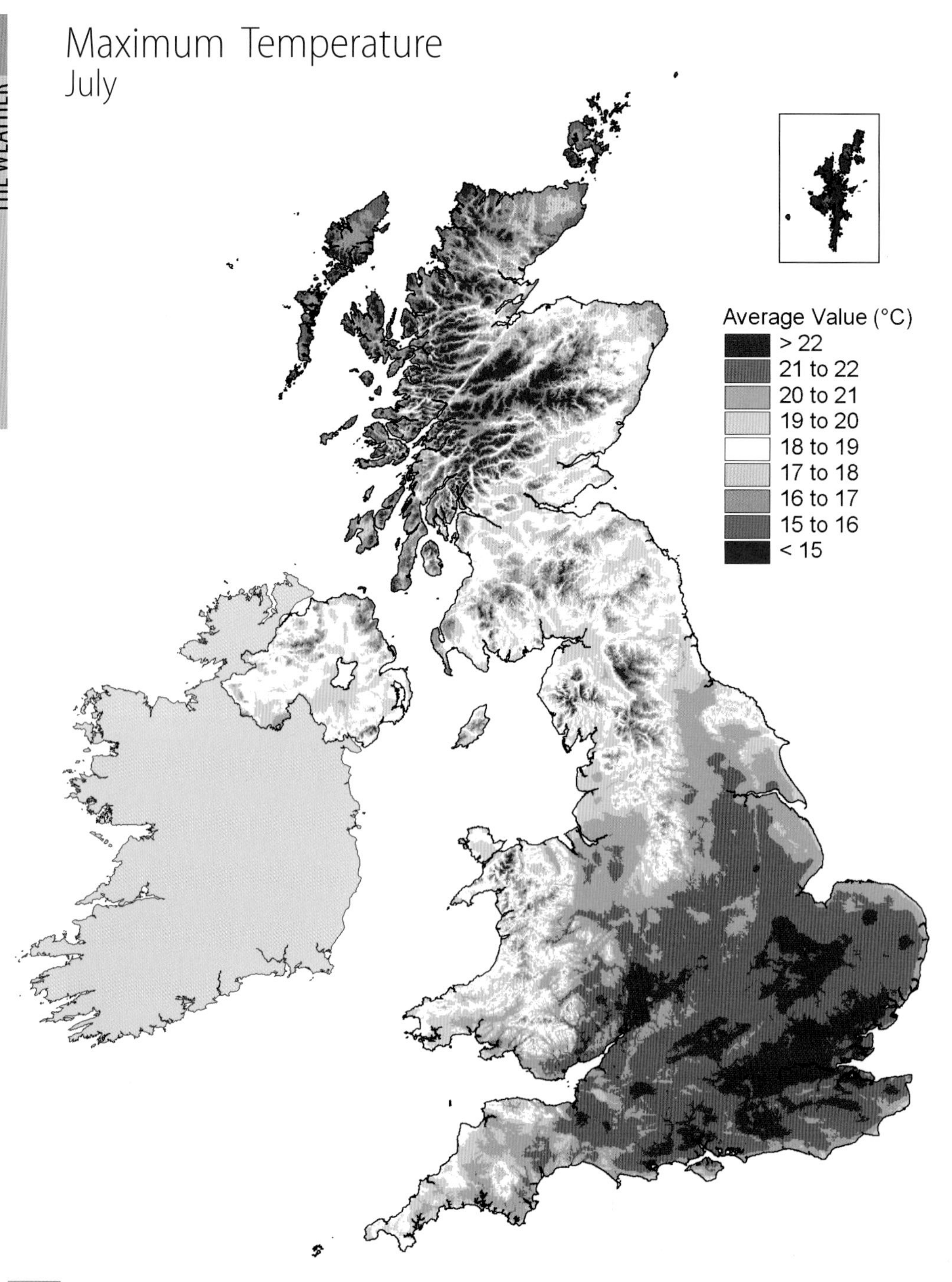
Average Value (°C)
> 22
21 to 22
20 to 21
19 to 20
18 to 19
17 to 18
16 to 17
15 to 16
< 15

Maximum Temperature
August

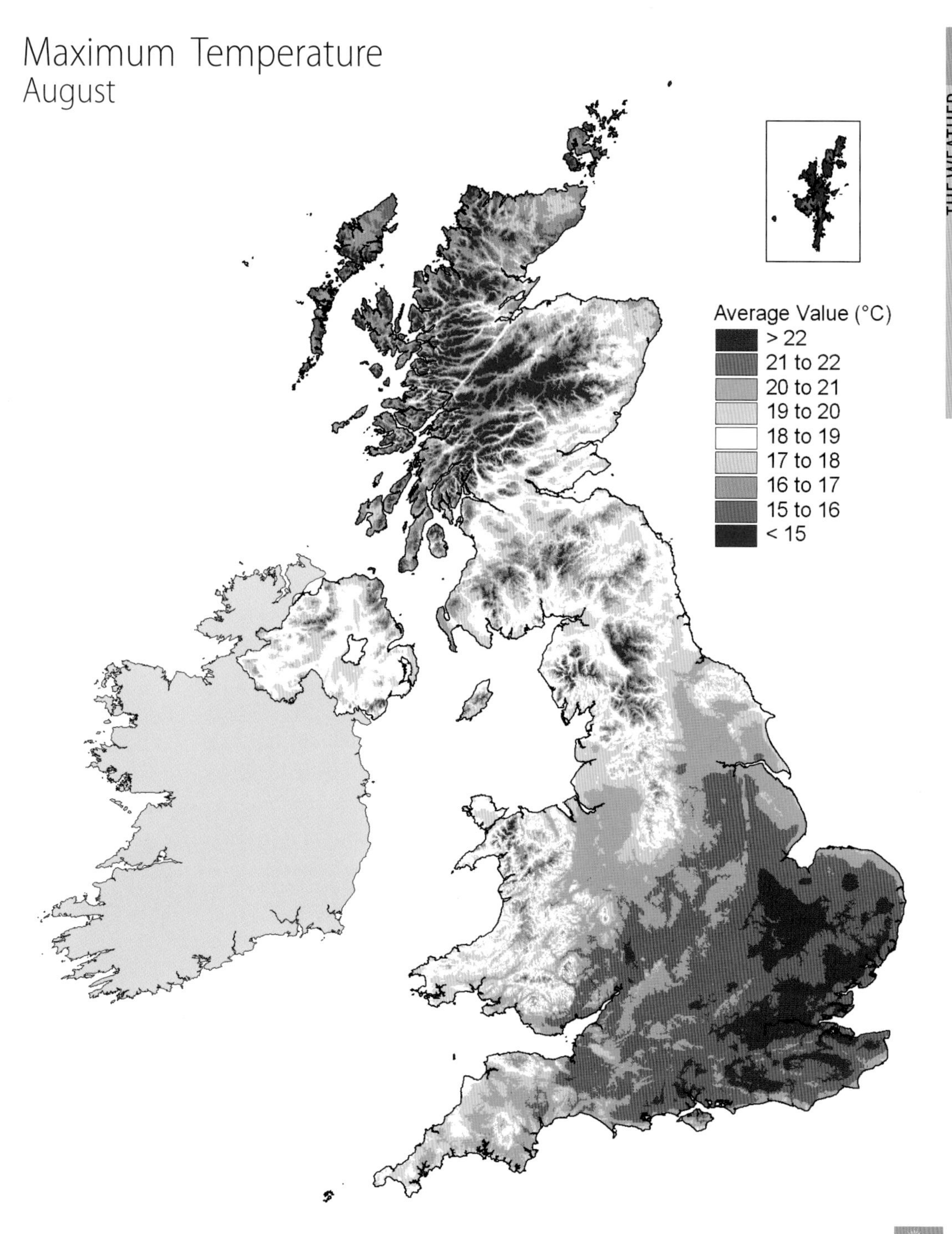
Average Value (°C)
> 22
21 to 22
20 to 21
19 to 20
18 to 19
17 to 18
16 to 17
15 to 16
< 15

Maximum Temperature
September

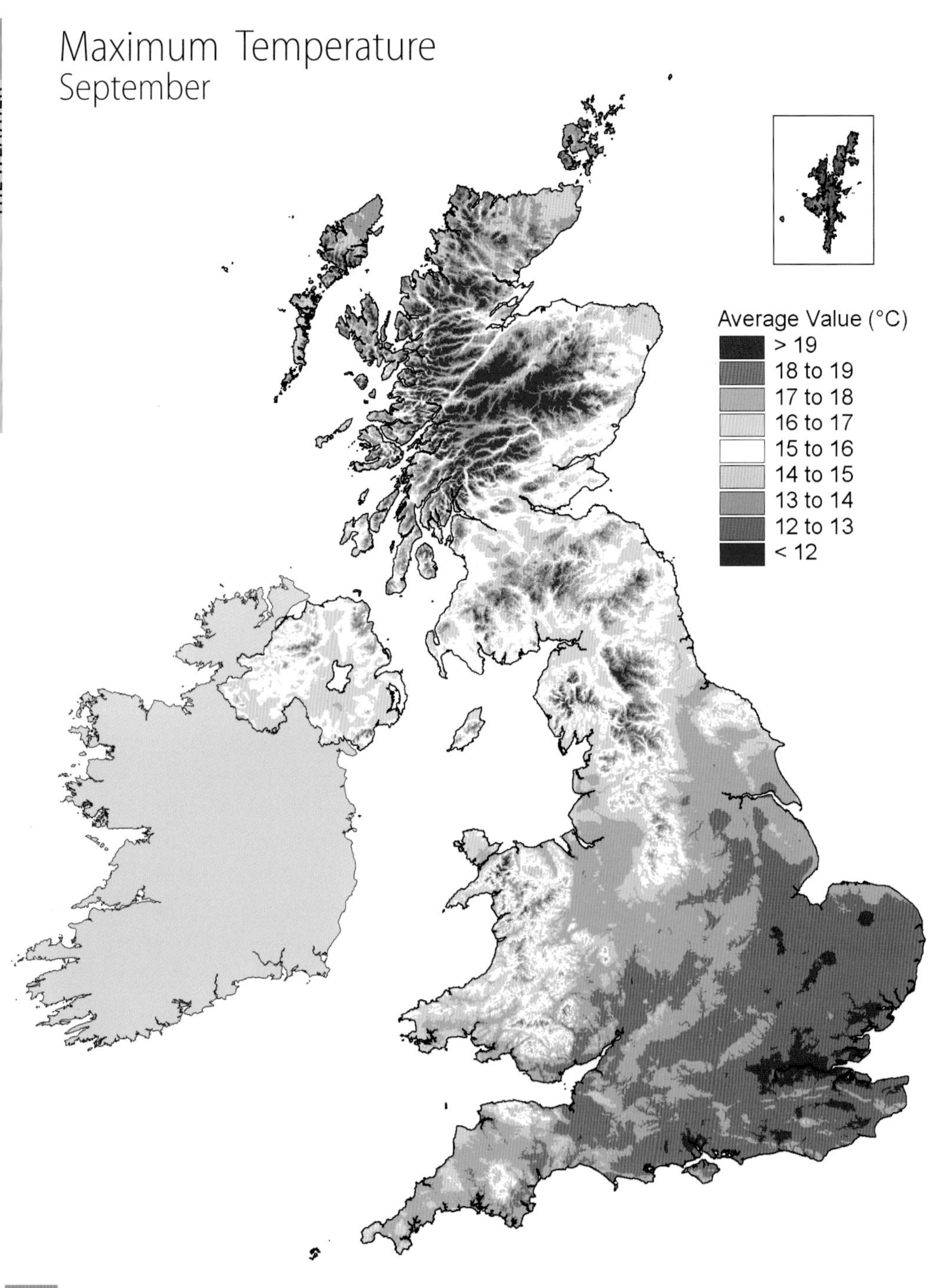
Average Value (°C)
> 19
18 to 19
17 to 18
16 to 17
15 to 16
14 to 15
13 to 14
12 to 13
< 12

Maximum Temperature
October

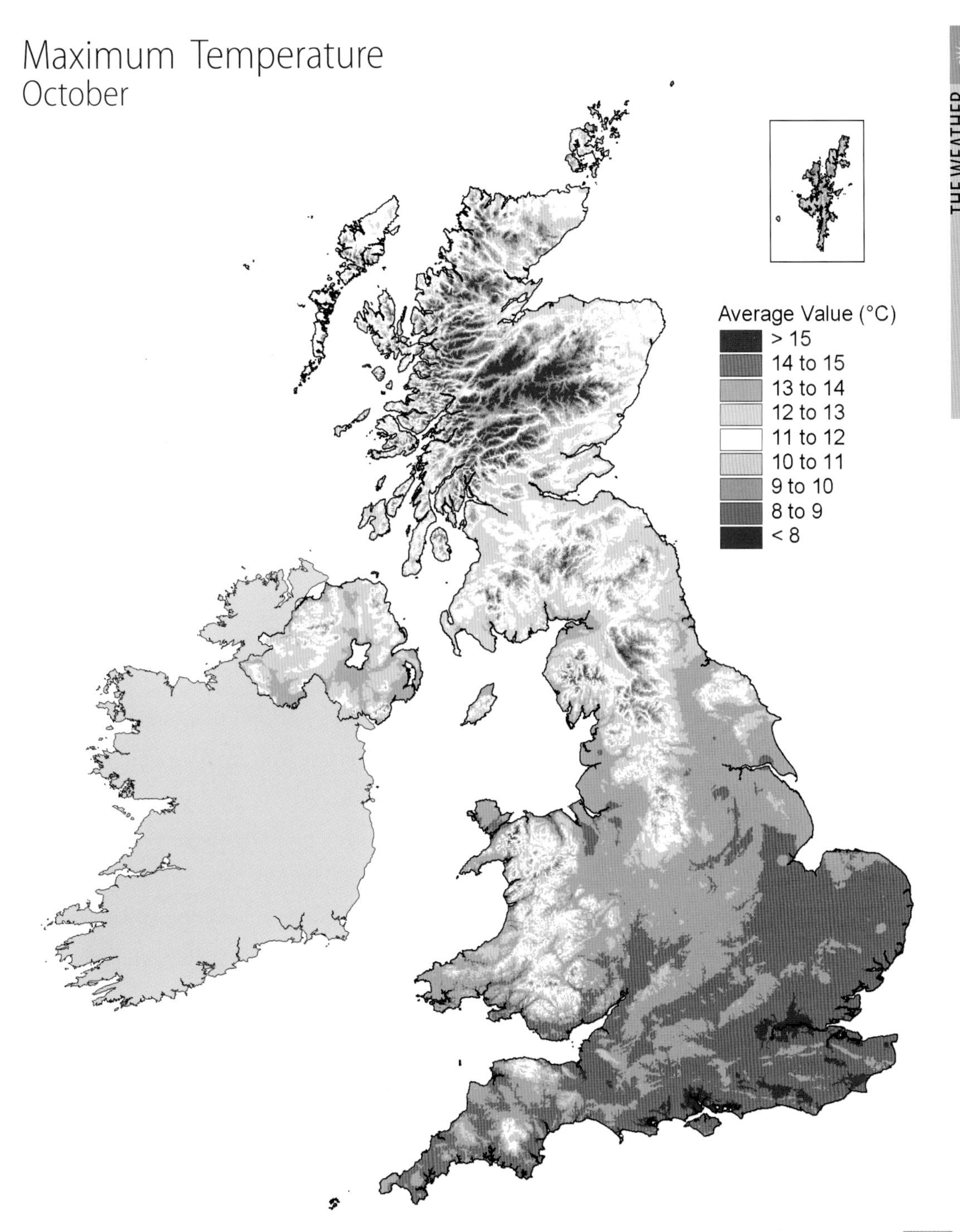
Average Value (°C)
> 15
14 to 15
13 to 14
12 to 13
11 to 12
10 to 11
9 to 10
8 to 9
< 8

Maximum Temperature
November

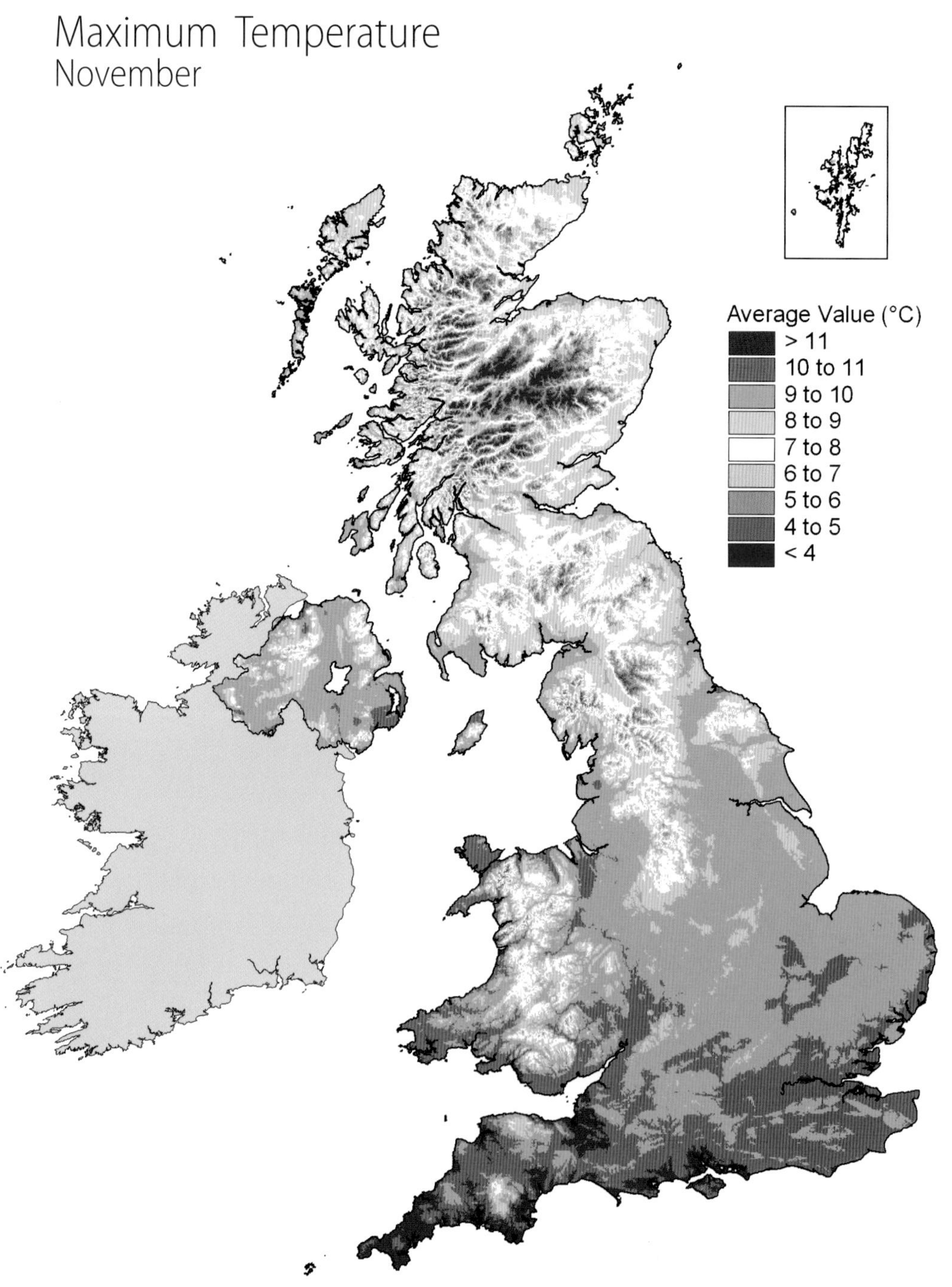
Average Value (°C)
> 11
10 to 11
9 to 10
8 to 9
7 to 8
6 to 7
5 to 6
4 to 5
< 4

Maximum Temperature
December

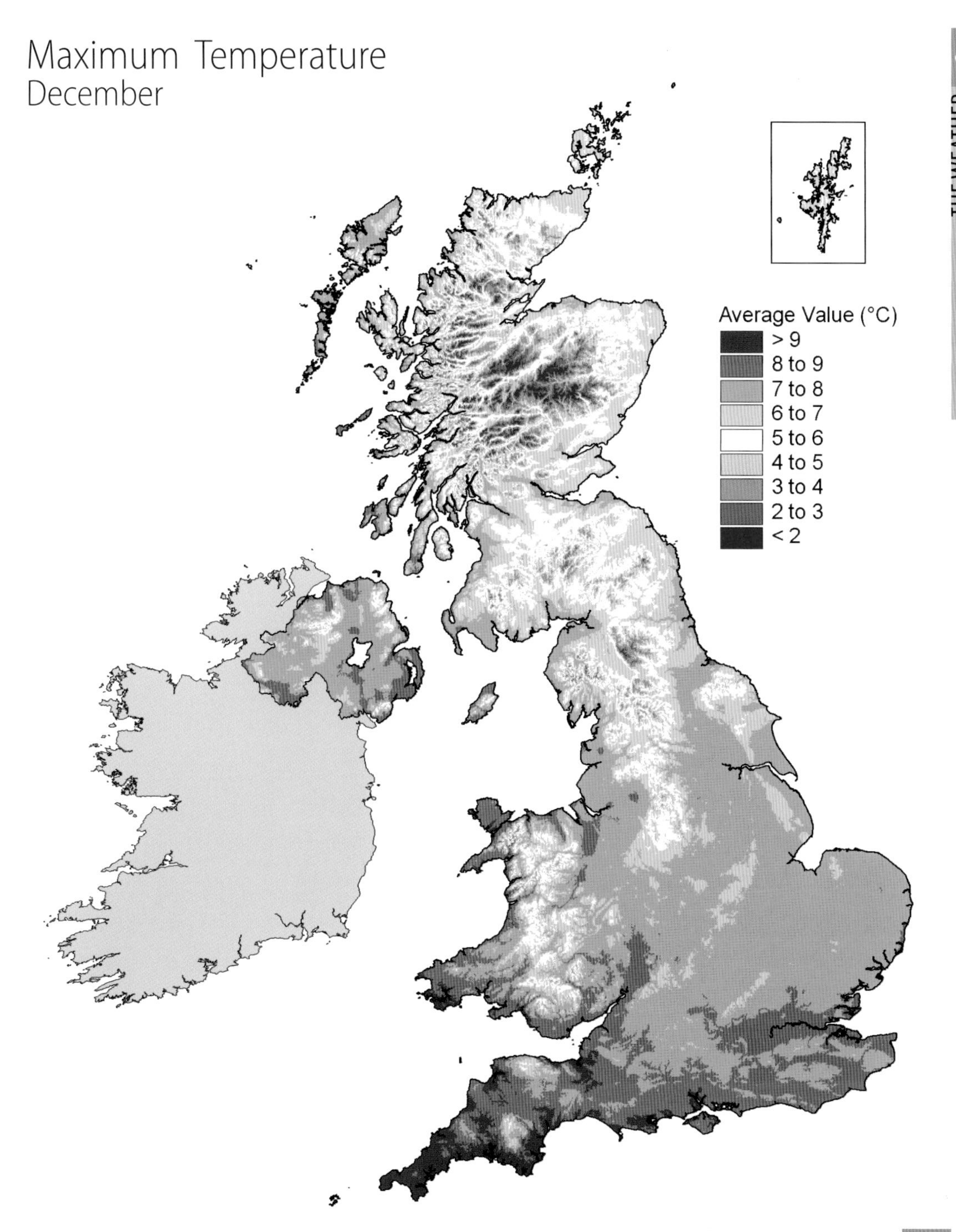
Average Value (°C)
> 9
8 to 9
7 to 8
6 to 7
5 to 6
4 to 5
3 to 4
2 to 3
< 2

Minimum Temperature

Month-by-month

Averages of night-time air temperatures reveal a regional pattern similar to that in the daytime, with the south and west mildest in winter and eastern areas favoured too in summer. However, the relatively warm local climate of urban and coastal areas is also apparent. In January and February, Cornwall again fares well, with averages close to 3°C, but London is not far behind. London and other cities in England are among the warmest spots in July and August.

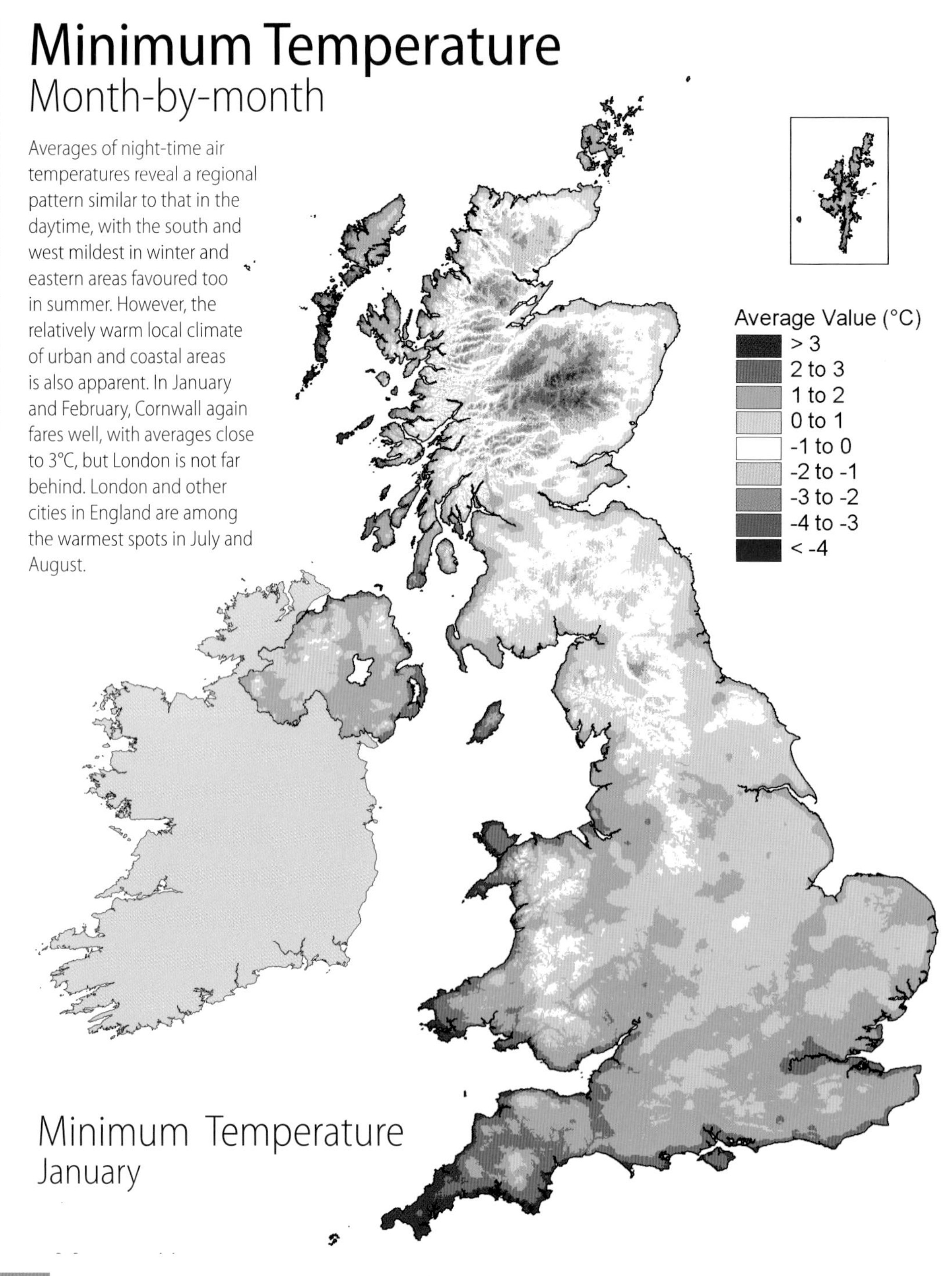
Average Value (°C)
> 3
2 to 3
1 to 2
0 to 1
-1 to 0
-2 to -1
-3 to -2
-4 to -3
< -4

Minimum Temperature
January

Minimum Temperature
February

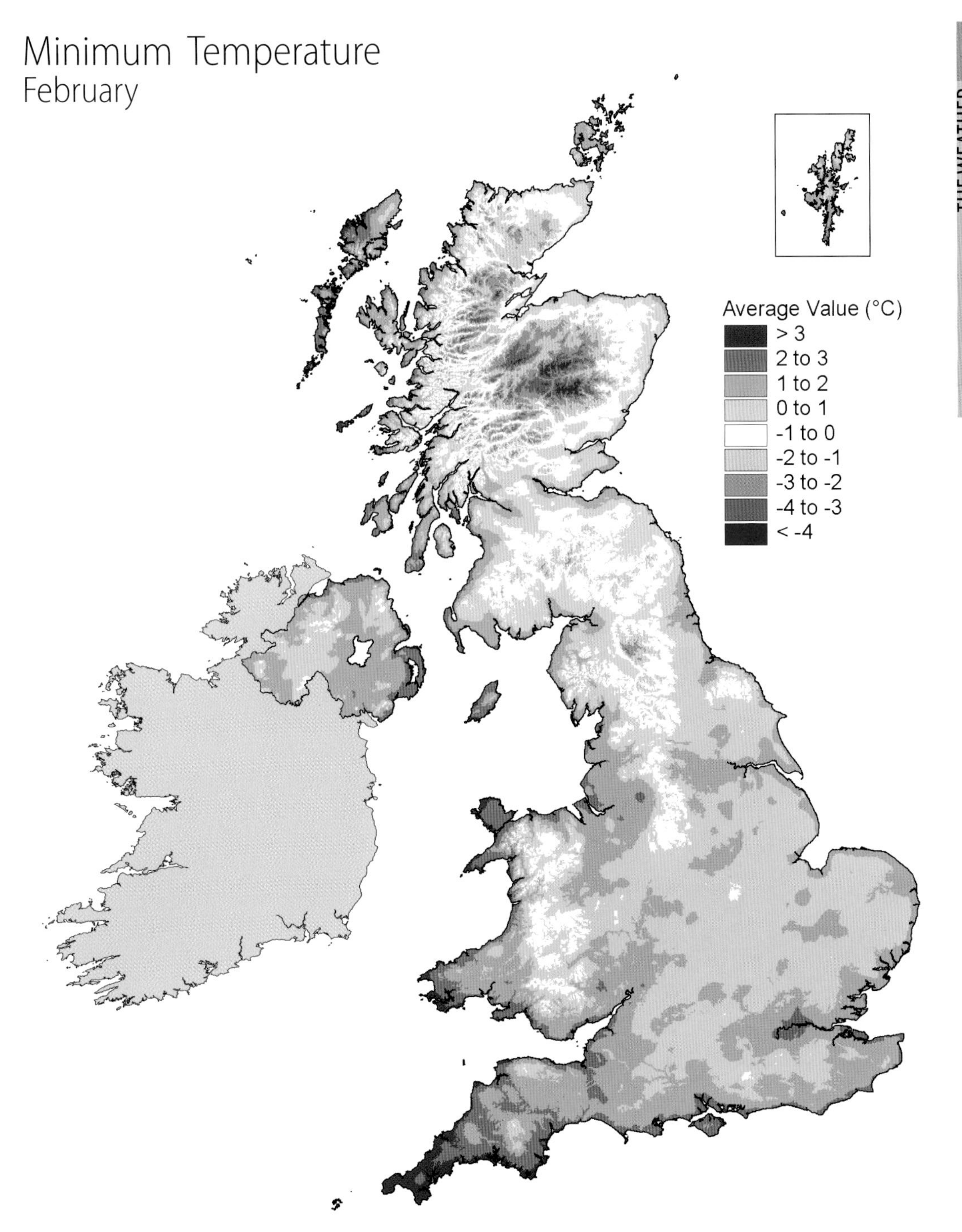

Average Value (°C)
> 3
2 to 3
1 to 2
0 to 1
-1 to 0
-2 to -1
-3 to -2
-4 to -3
< -4

Minimum Temperature
March

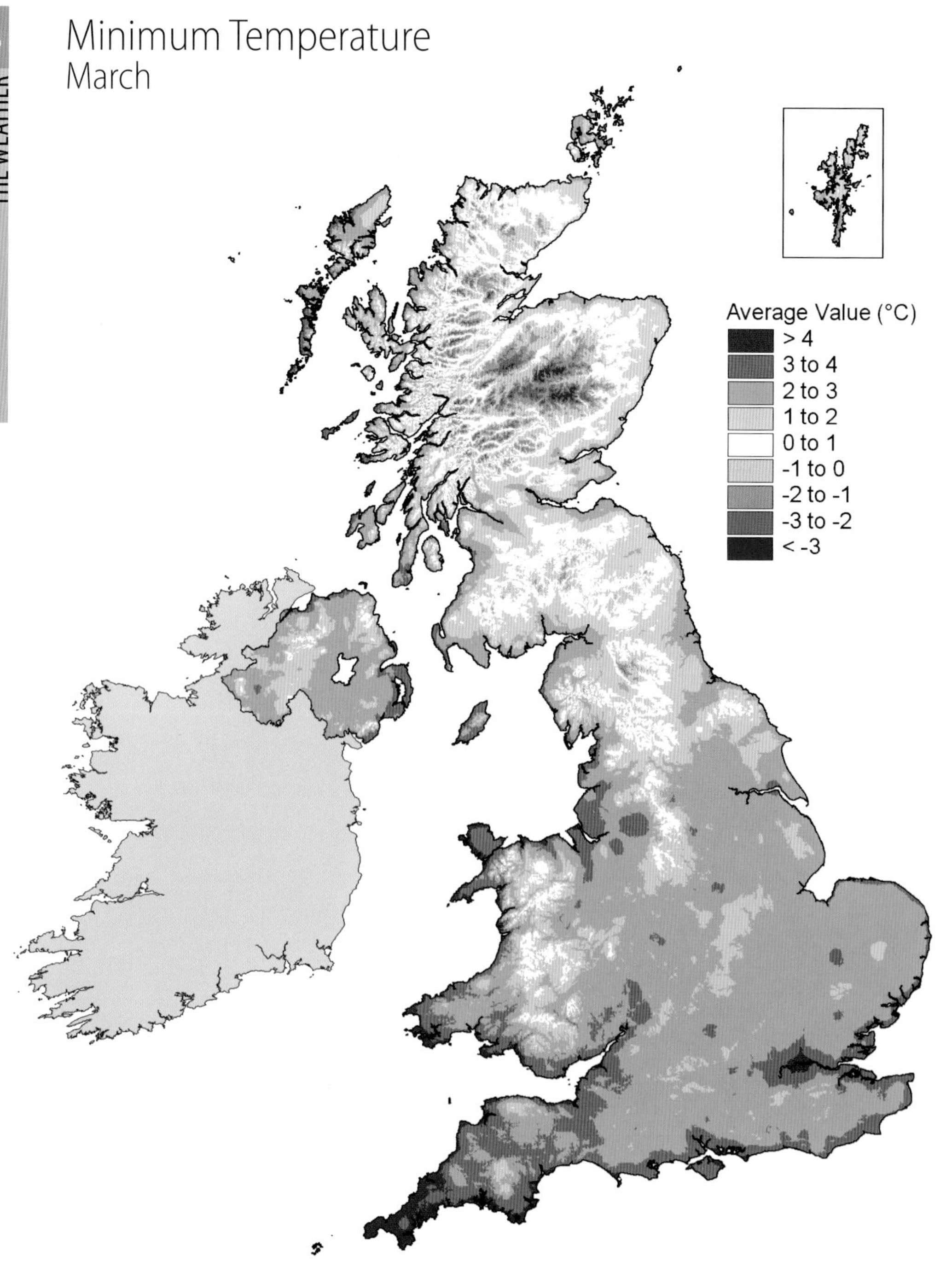

Average Value (°C)
> 4
3 to 4
2 to 3
1 to 2
0 to 1
-1 to 0
-2 to -1
-3 to -2
< -3

Minimum Temperature
April

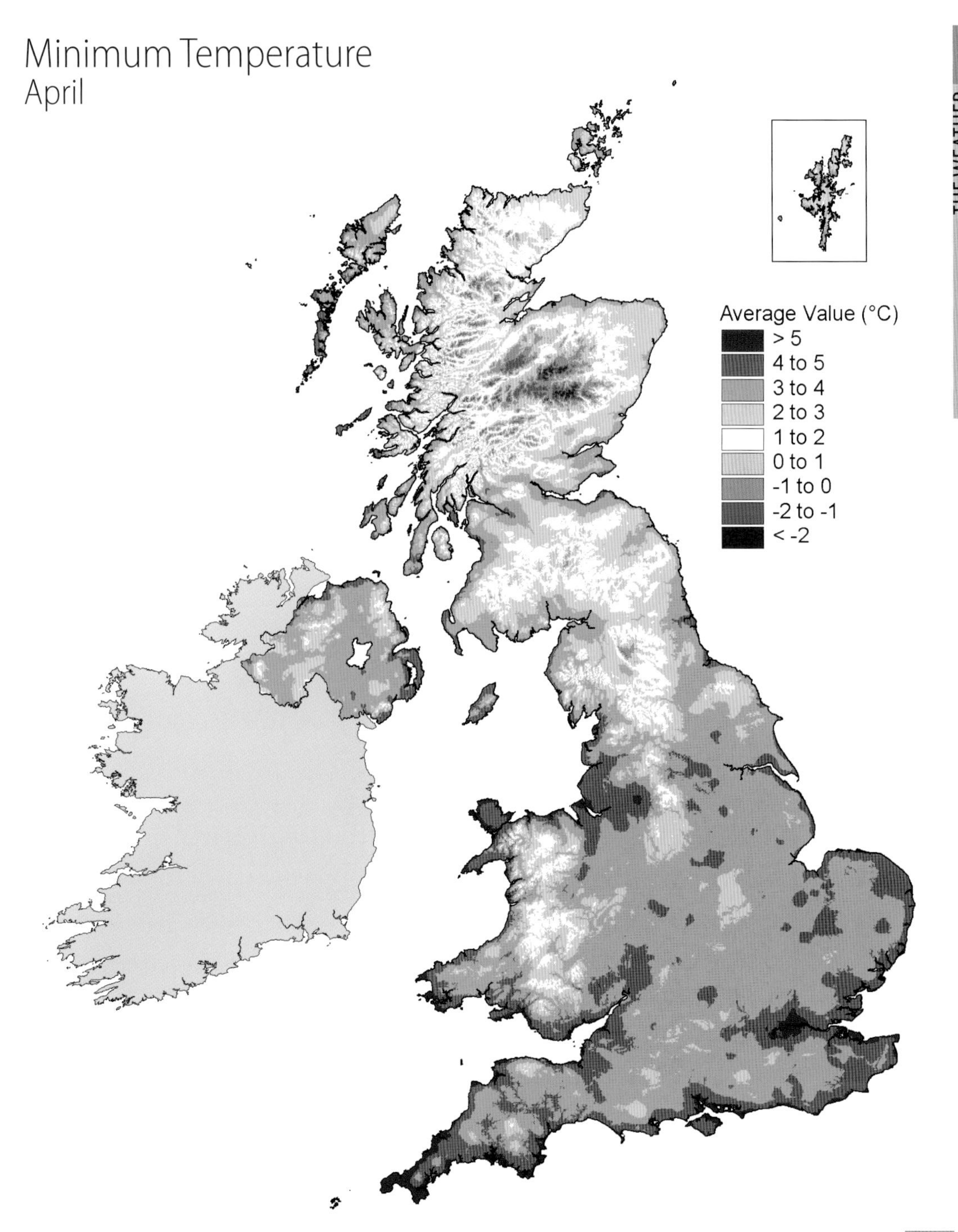
Average Value (°C)
> 5
4 to 5
3 to 4
2 to 3
1 to 2
0 to 1
-1 to 0
-2 to -1
< -2

Minimum Temperature
May

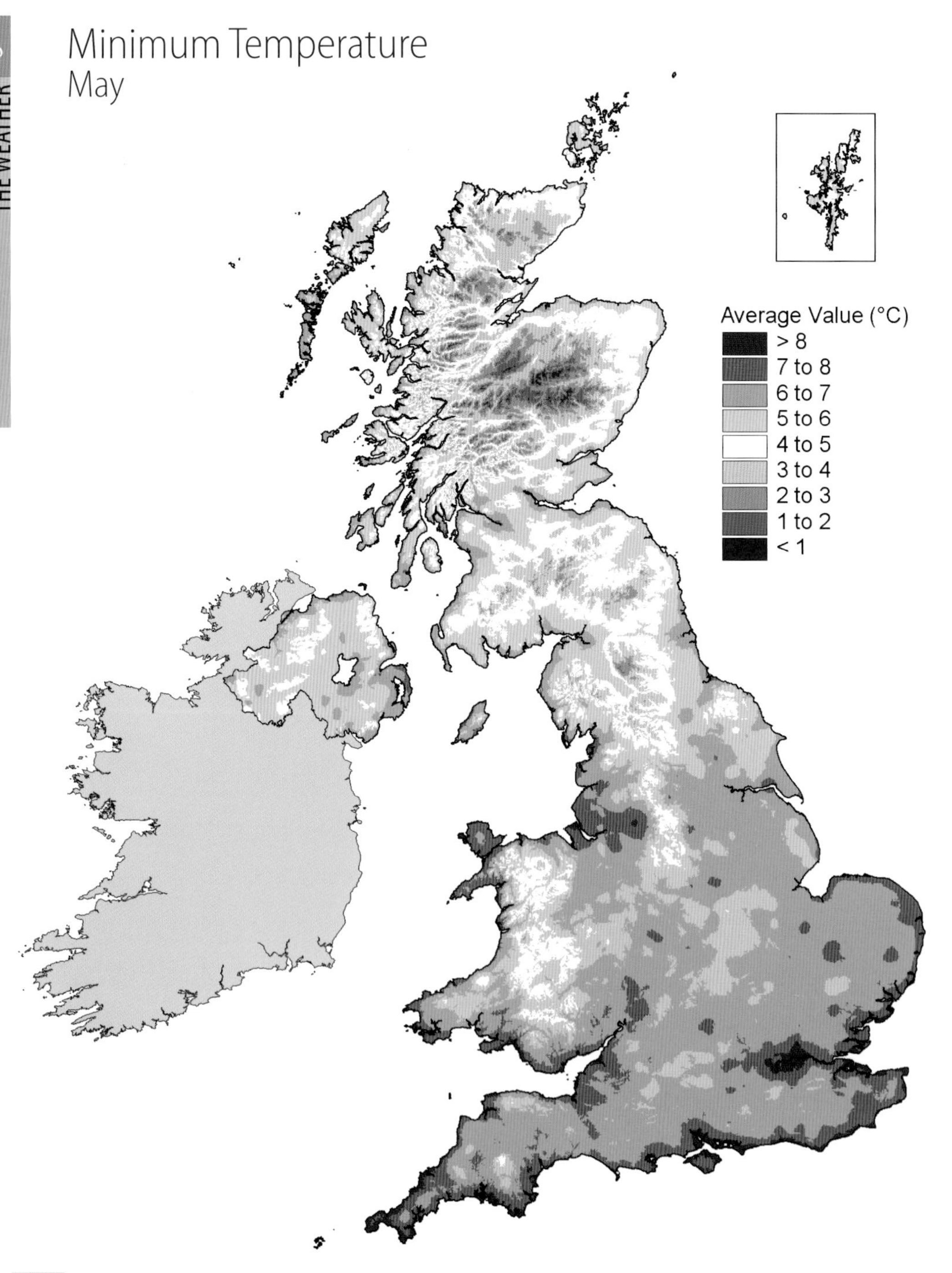
Average Value (°C)
> 8
7 to 8
6 to 7
5 to 6
4 to 5
3 to 4
2 to 3
1 to 2
< 1

Minimum Temperature
June

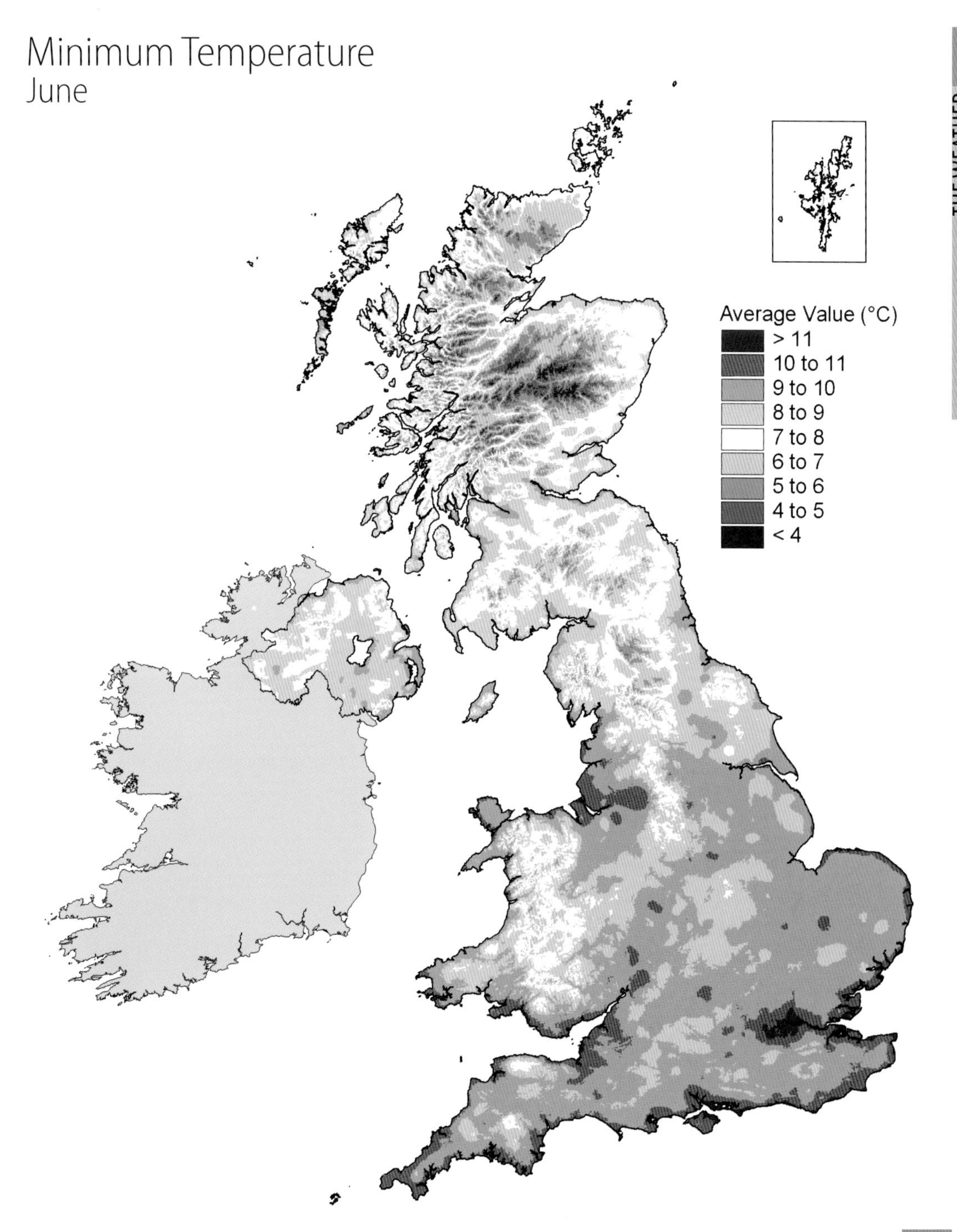
Average Value (°C)
> 11
10 to 11
9 to 10
8 to 9
7 to 8
6 to 7
5 to 6
4 to 5
< 4

Minimum Temperature
July

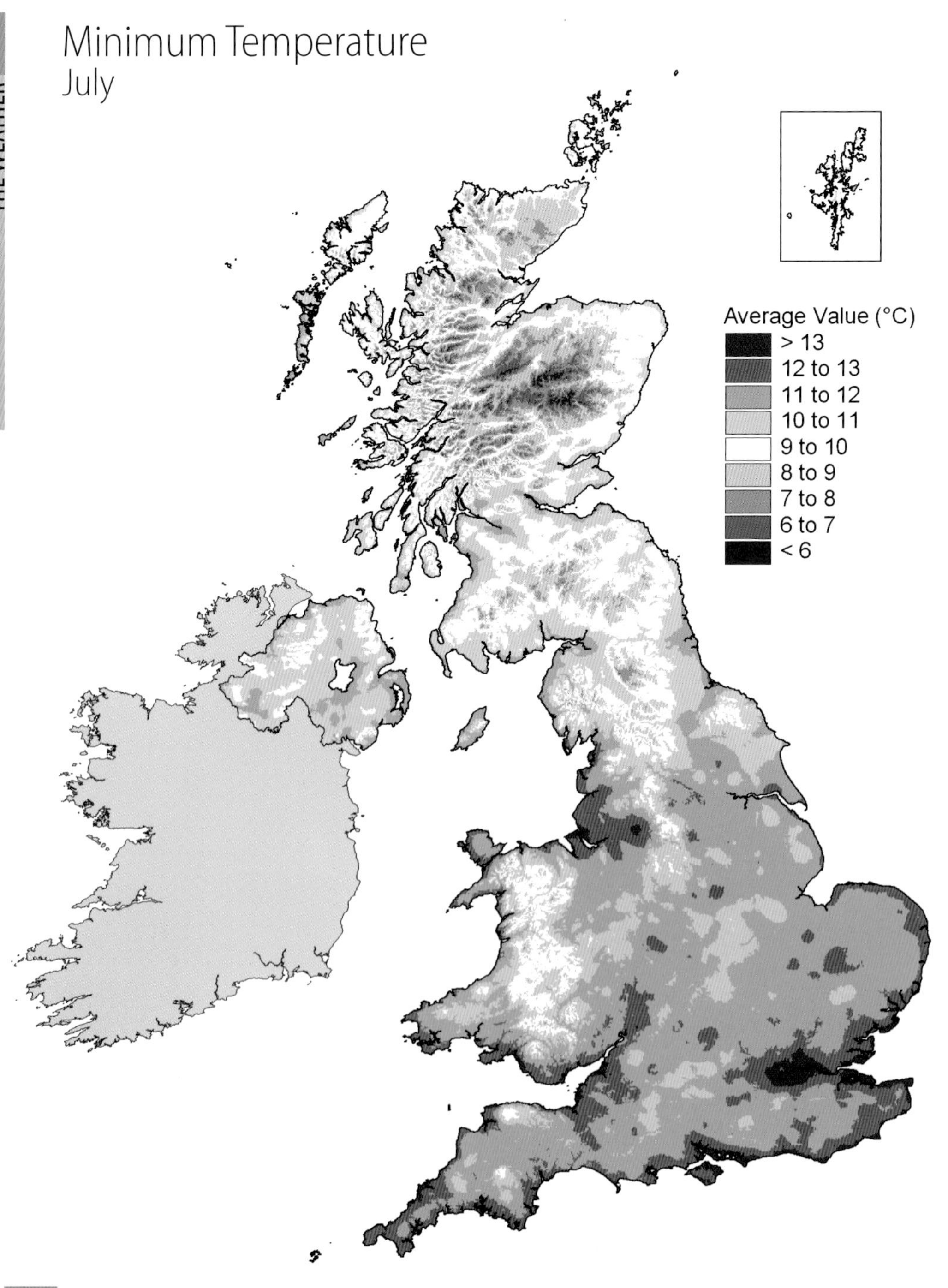
Average Value (°C)
> 13
12 to 13
11 to 12
10 to 11
9 to 10
8 to 9
7 to 8
6 to 7
< 6

Minimum Temperature
August

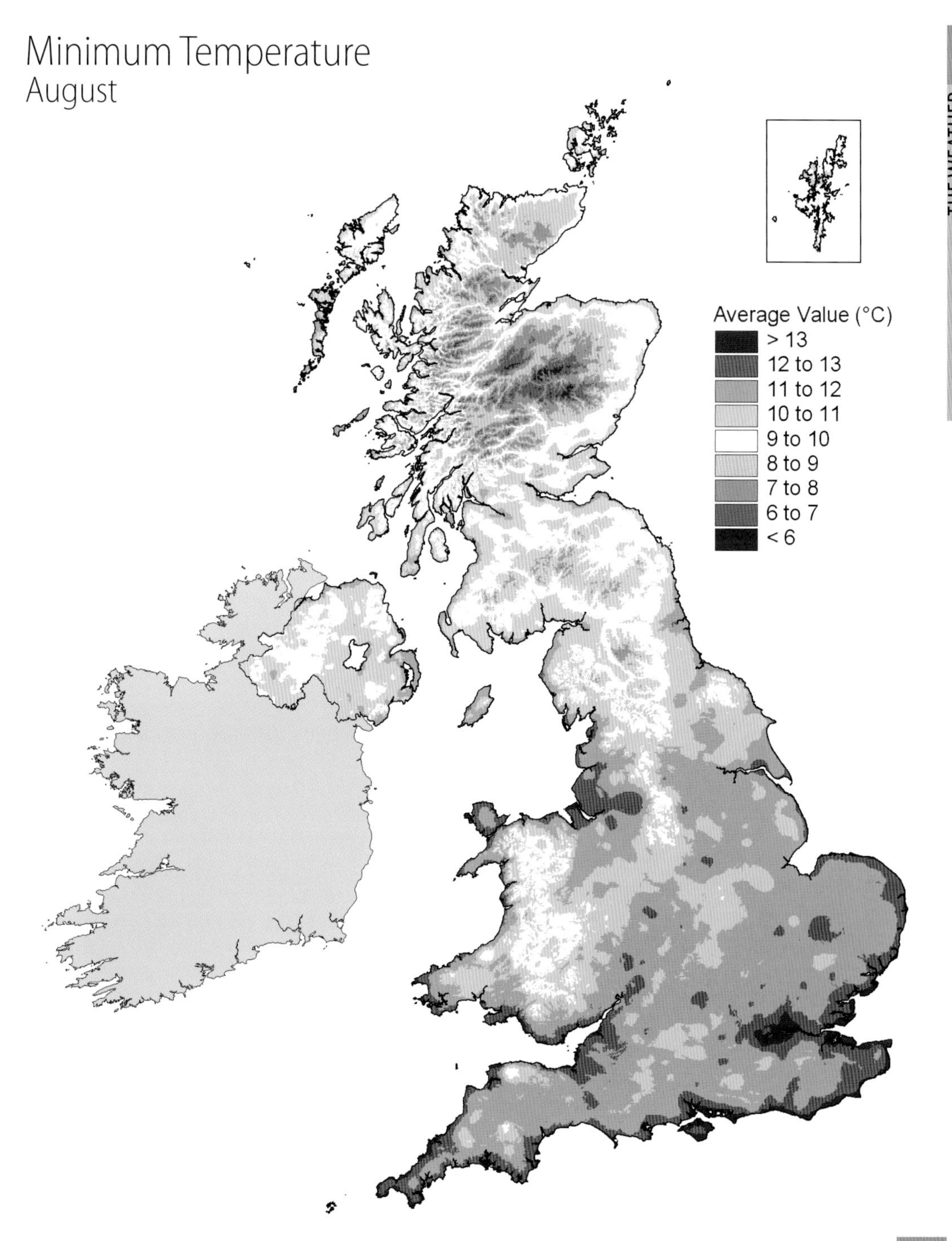
Average Value (°C)
> 13
12 to 13
11 to 12
10 to 11
9 to 10
8 to 9
7 to 8
6 to 7
< 6

Minimum Temperature
September

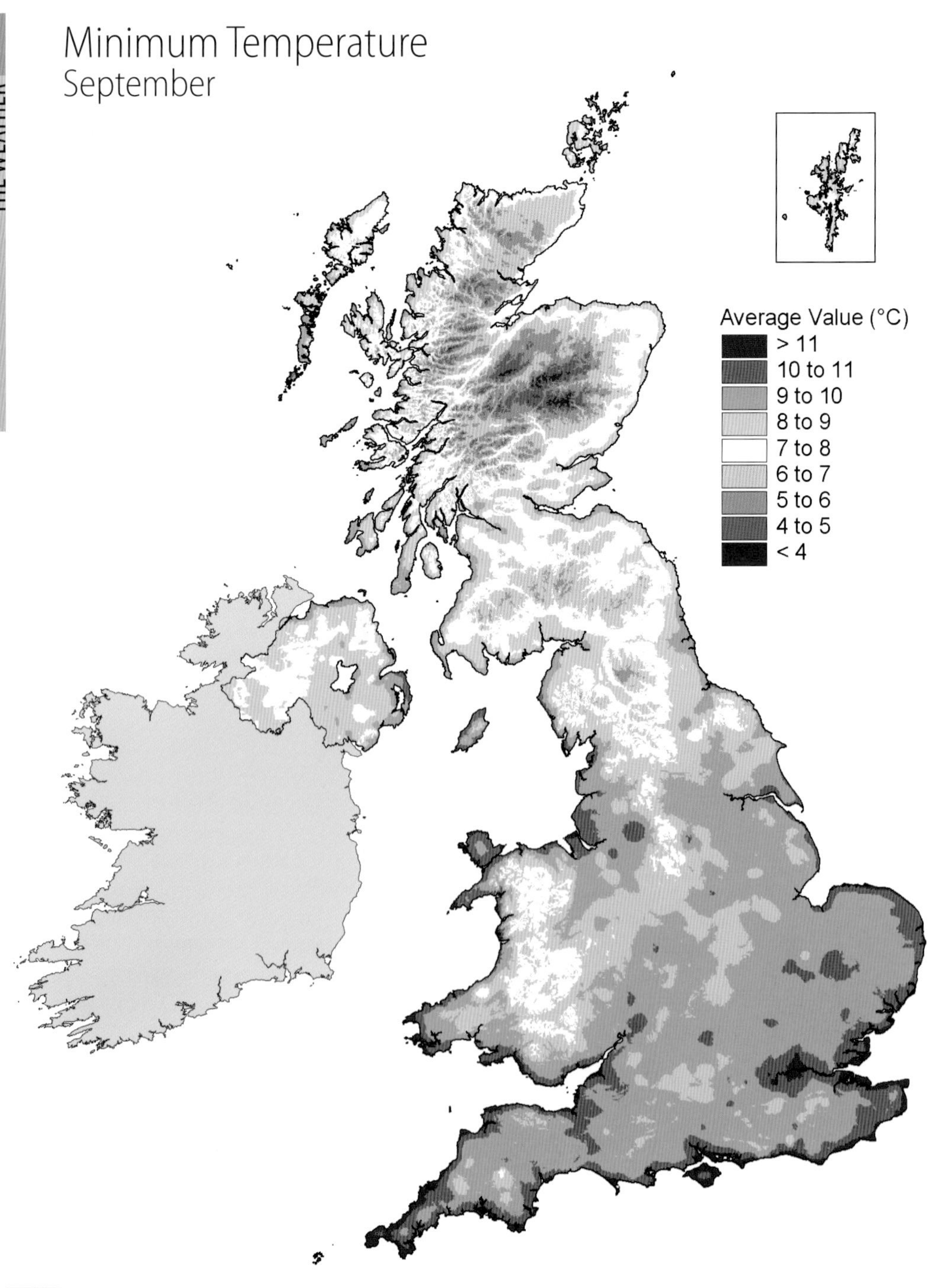
Average Value (°C)
> 11
10 to 11
9 to 10
8 to 9
7 to 8
6 to 7
5 to 6
4 to 5
< 4

Minimum Temperature
October

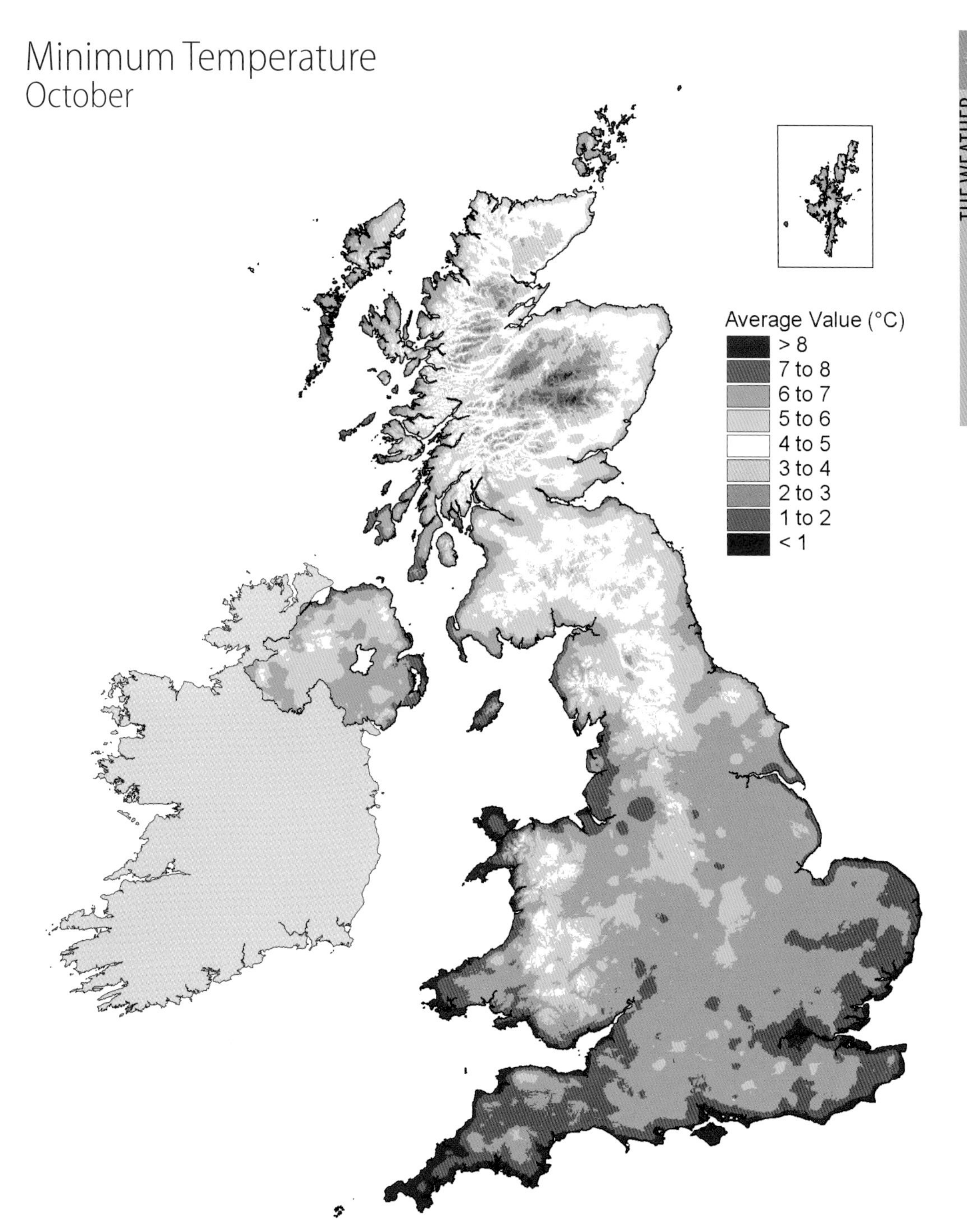
Average Value (°C)
> 8
7 to 8
6 to 7
5 to 6
4 to 5
3 to 4
2 to 3
1 to 2
< 1

Minimum Temperature
November

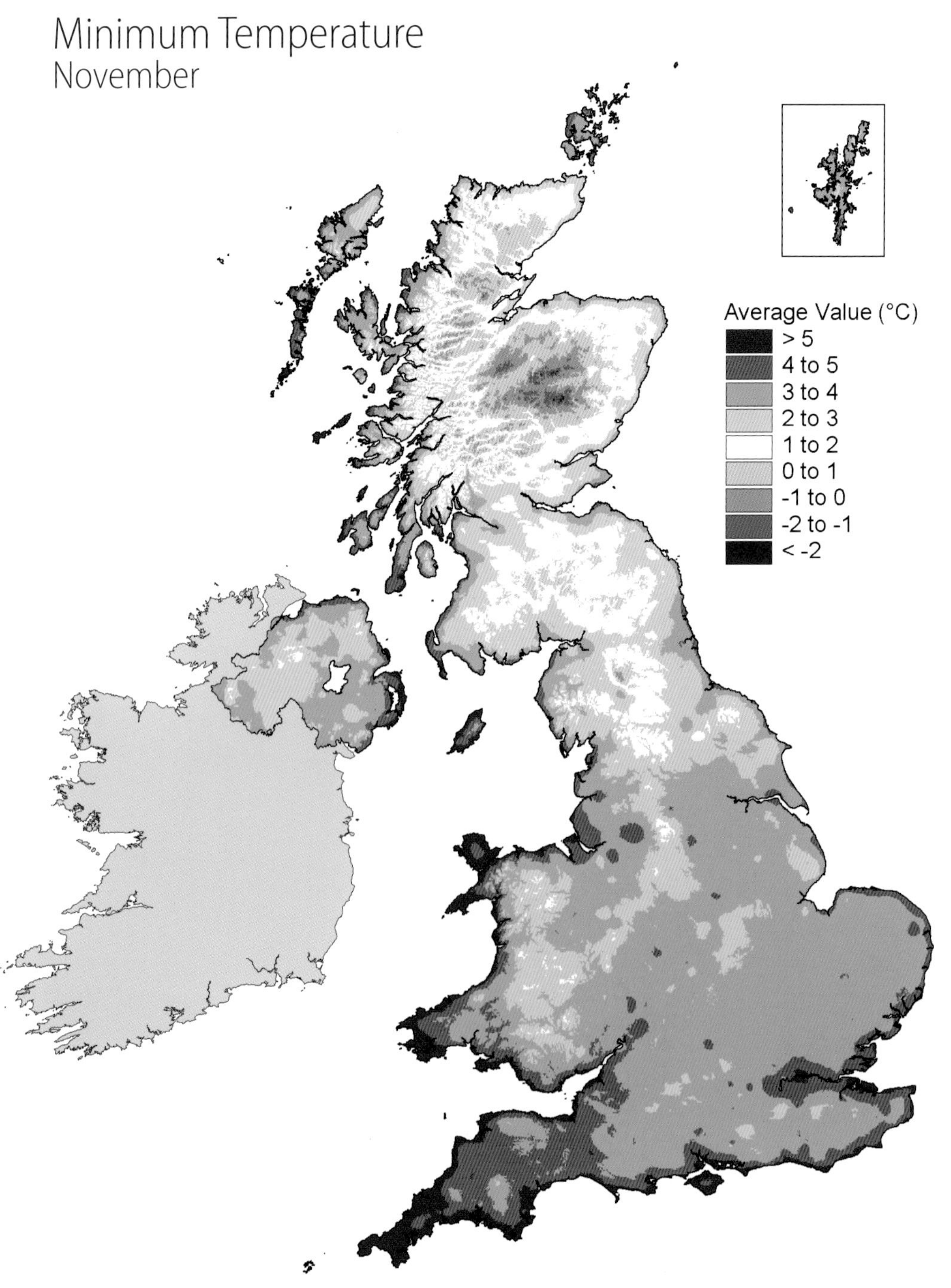
Average Value (°C)
> 5
4 to 5
3 to 4
2 to 3
1 to 2
0 to 1
-1 to 0
-2 to -1
< -2

Minimum Temperature
December

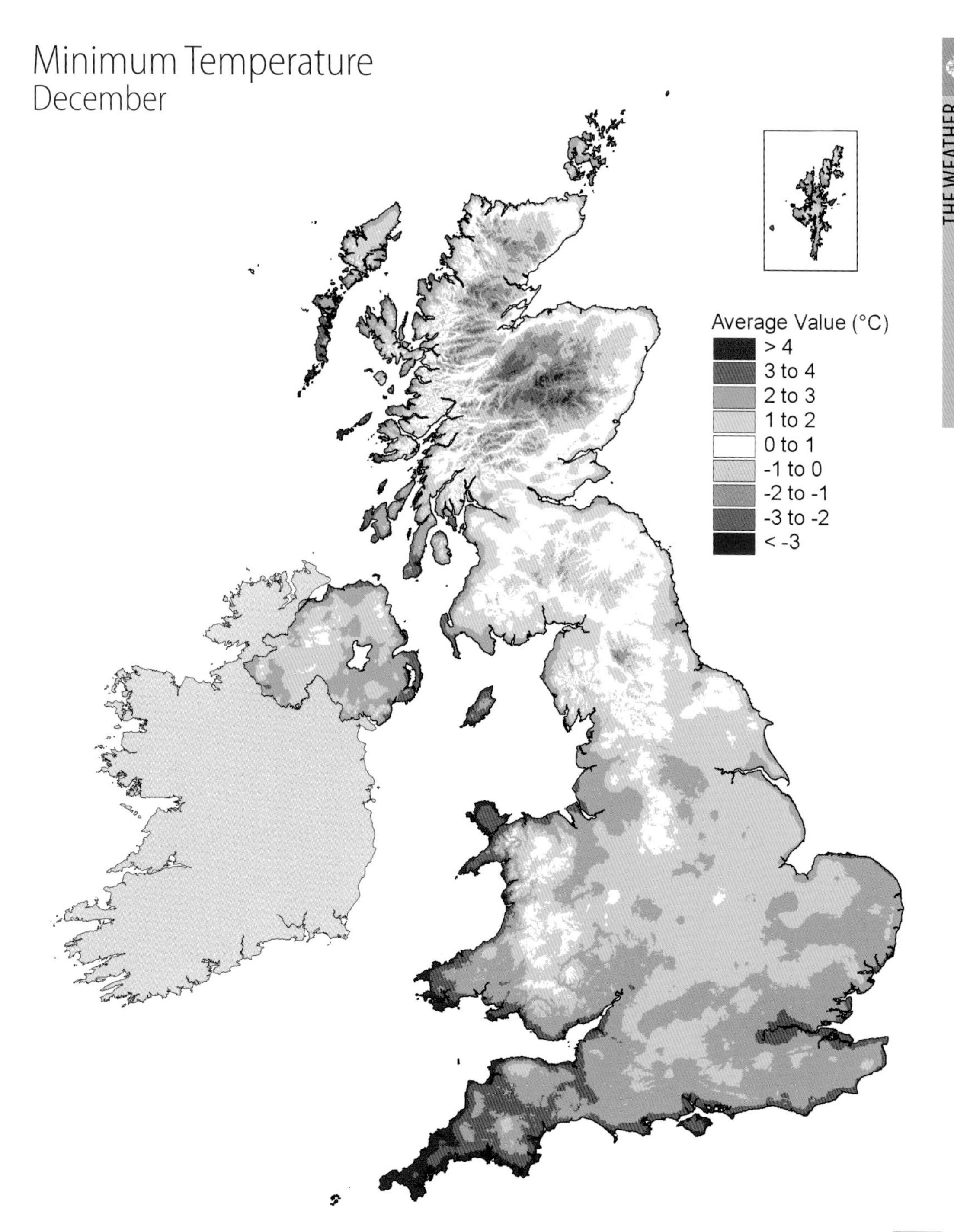
Average Value (°C)
> 4
3 to 4
2 to 3
1 to 2
0 to 1
-1 to 0
-2 to -1
-3 to -2
< -3

Wind Speed
Month-by-month

The strength of the wind is closely linked to the passage of Atlantic weather systems so the windiest places tend to be on the coastlines and high ground of the west and north. In contrast, places well inland and to the lee of high ground, such as the West Midlands, are much more sheltered. The months of November to February usually experience the highest wind speeds, while June to August have the lightest winds. Although the prevailing wind direction is south-westerly, May and June often bring spells of north-easterly winds.

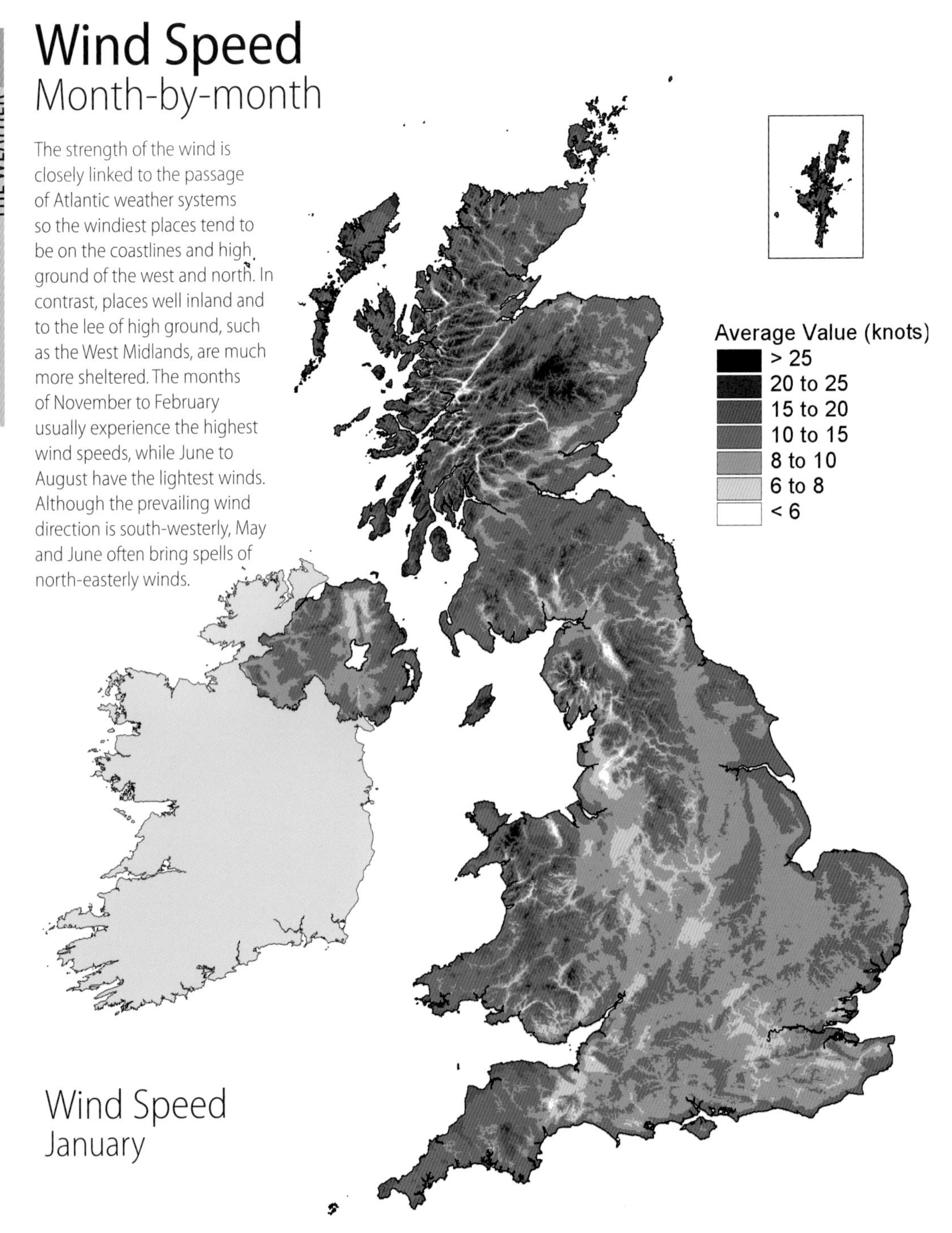
Average Value (knots)
> 25
20 to 25
15 to 20
10 to 15
8 to 10
6 to 8
< 6

Wind Speed
January

Wind Speed
February

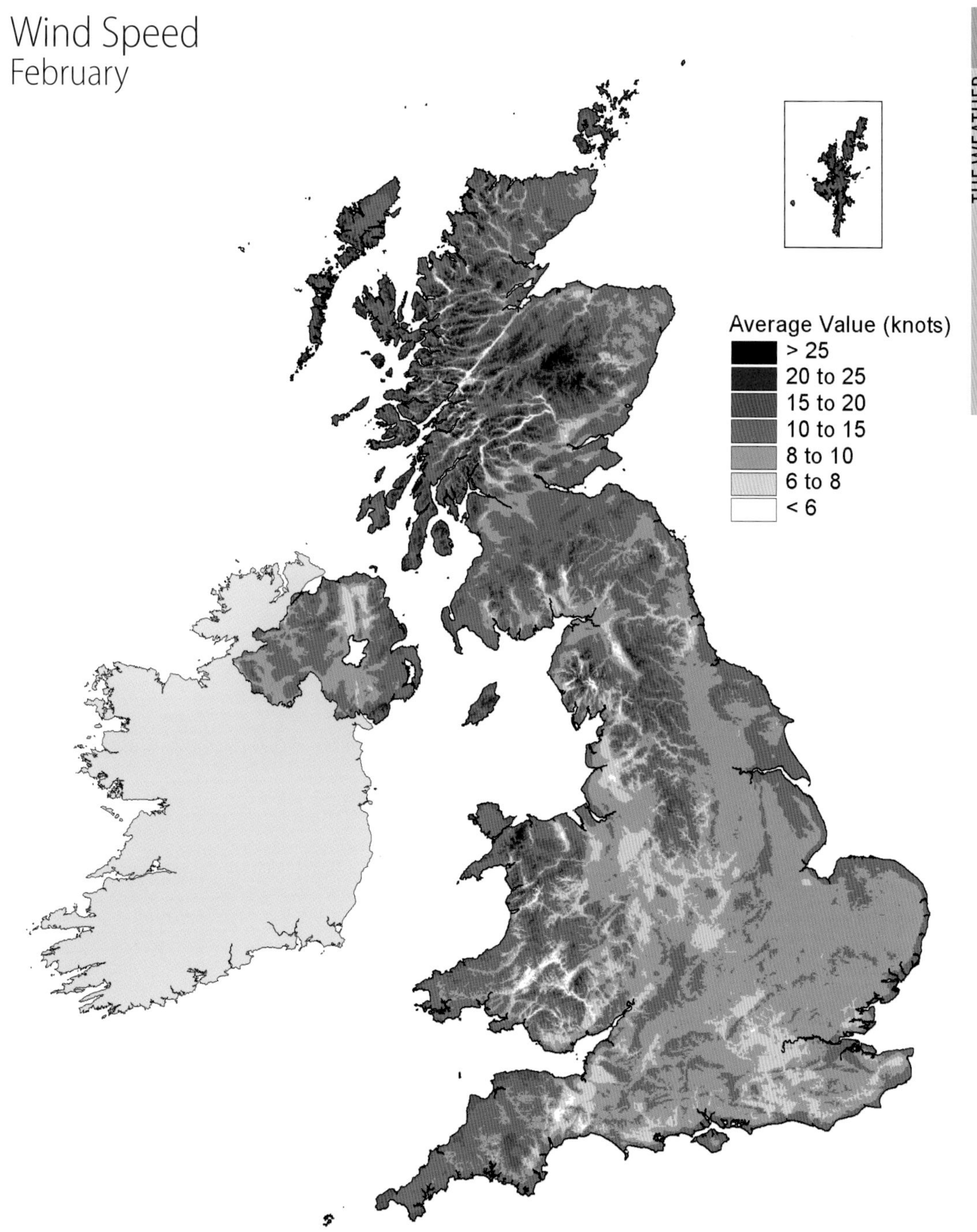
Average Value (knots)
> 25
20 to 25
15 to 20
10 to 15
8 to 10
6 to 8
< 6

Wind Speed
March

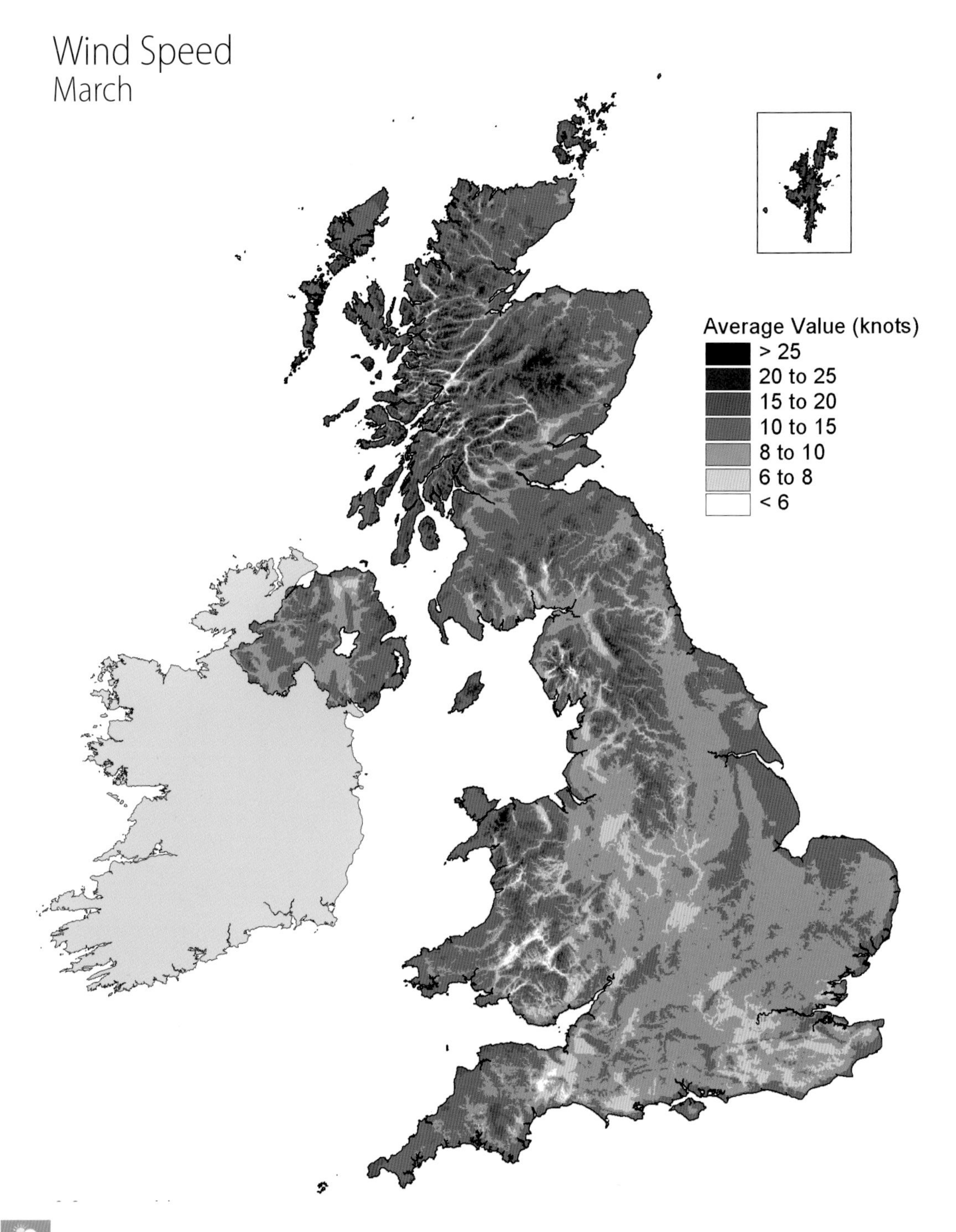

Average Value (knots)
> 25
20 to 25
15 to 20
10 to 15
8 to 10
6 to 8
< 6

Wind Speed
April

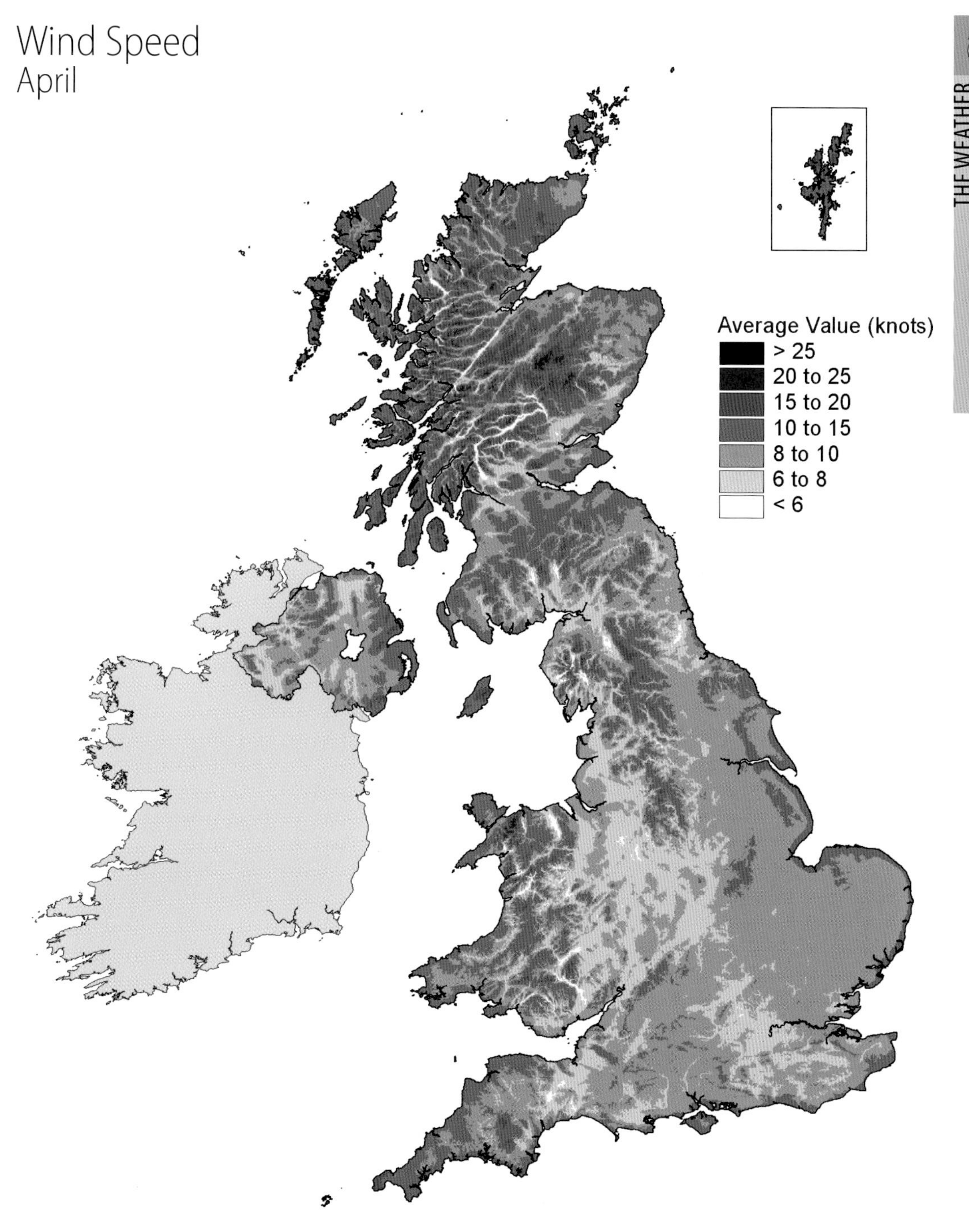
Average Value (knots)
> 25
20 to 25
15 to 20
10 to 15
8 to 10
6 to 8
< 6

Wind Speed
May

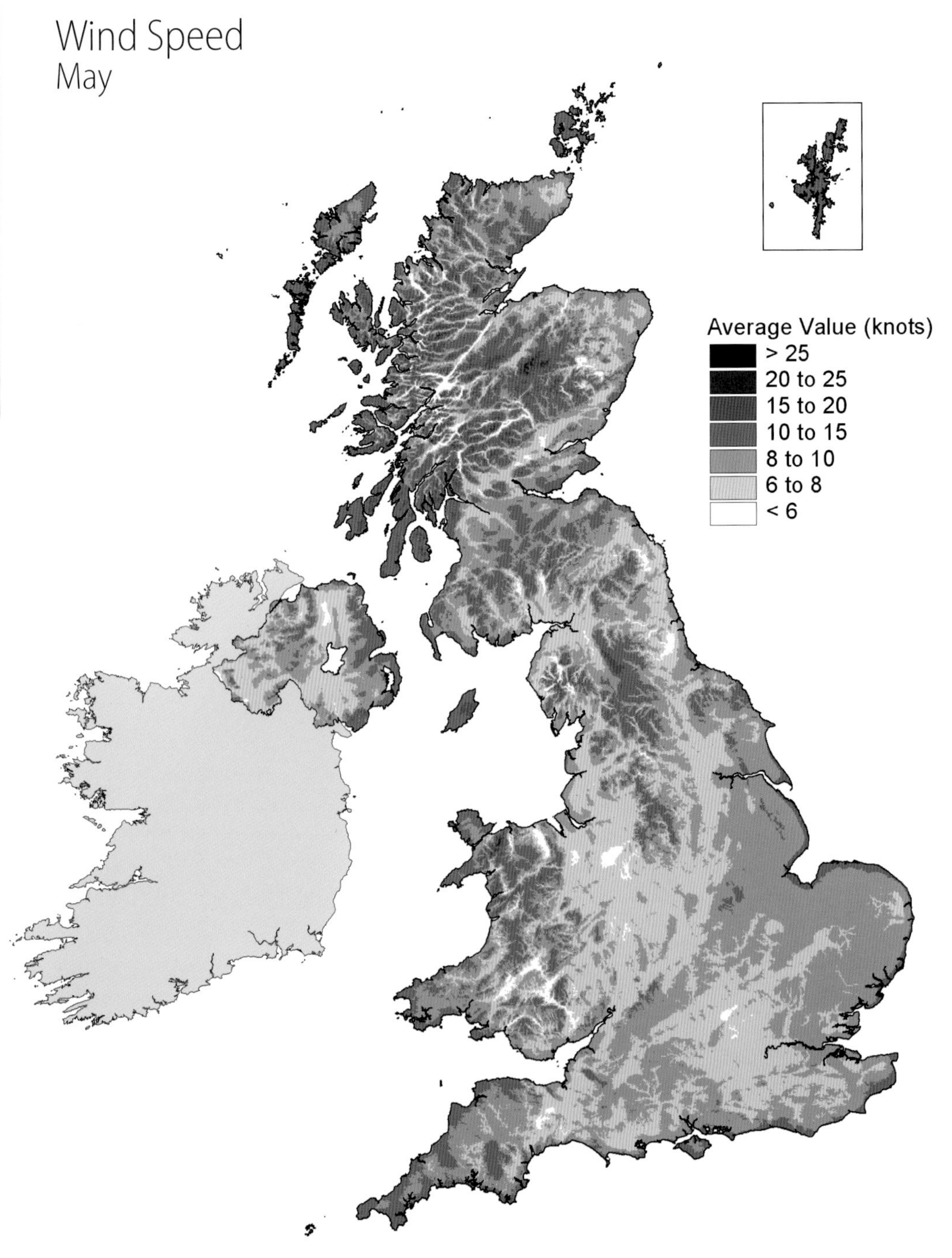
Average Value (knots)
> 25
20 to 25
15 to 20
10 to 15
8 to 10
6 to 8
< 6

Wind Speed
June

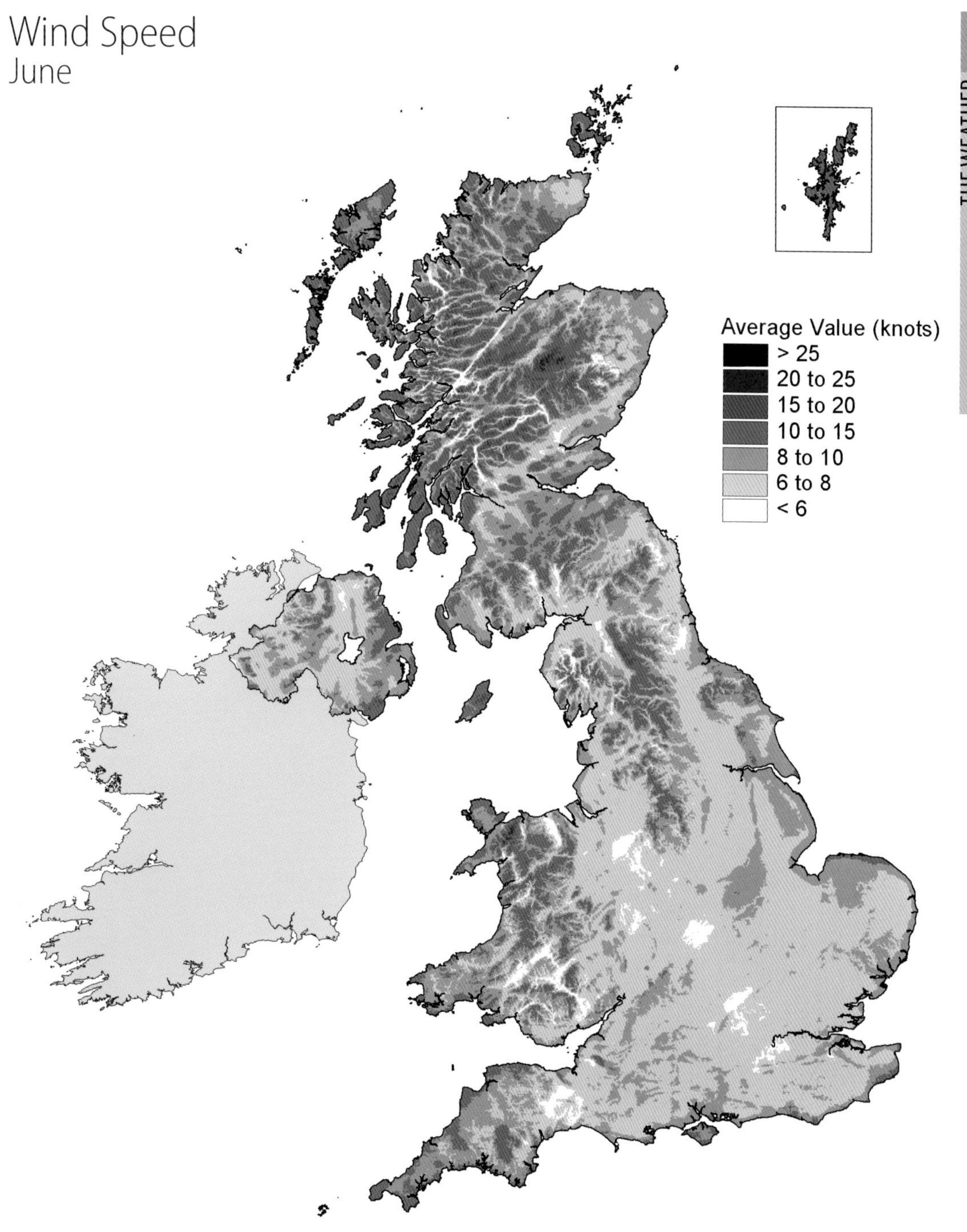
Average Value (knots)
> 25
20 to 25
15 to 20
10 to 15
8 to 10
6 to 8
< 6

Wind Speed
July

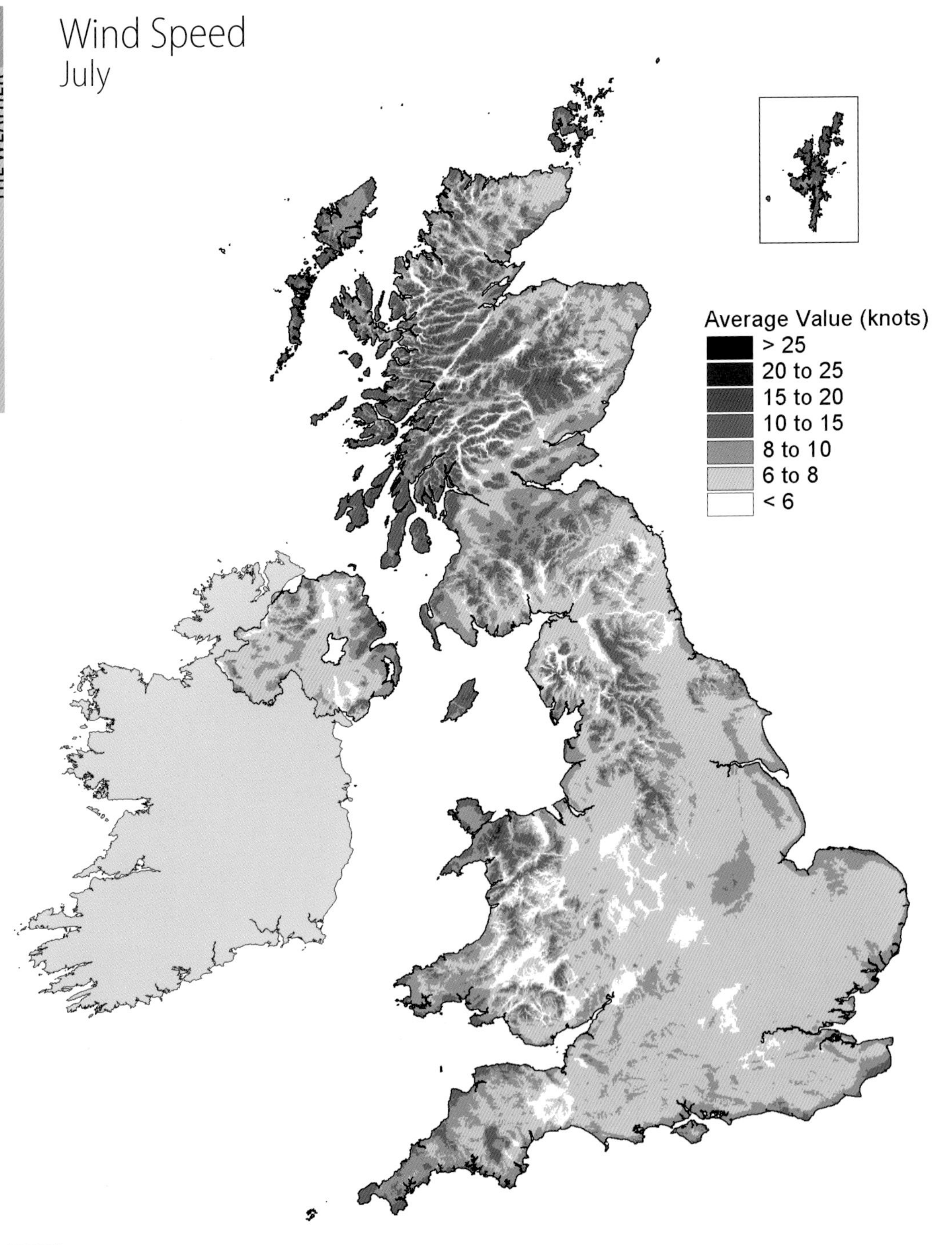
Average Value (knots)
> 25
20 to 25
15 to 20
10 to 15
8 to 10
6 to 8
< 6

Wind Speed
August

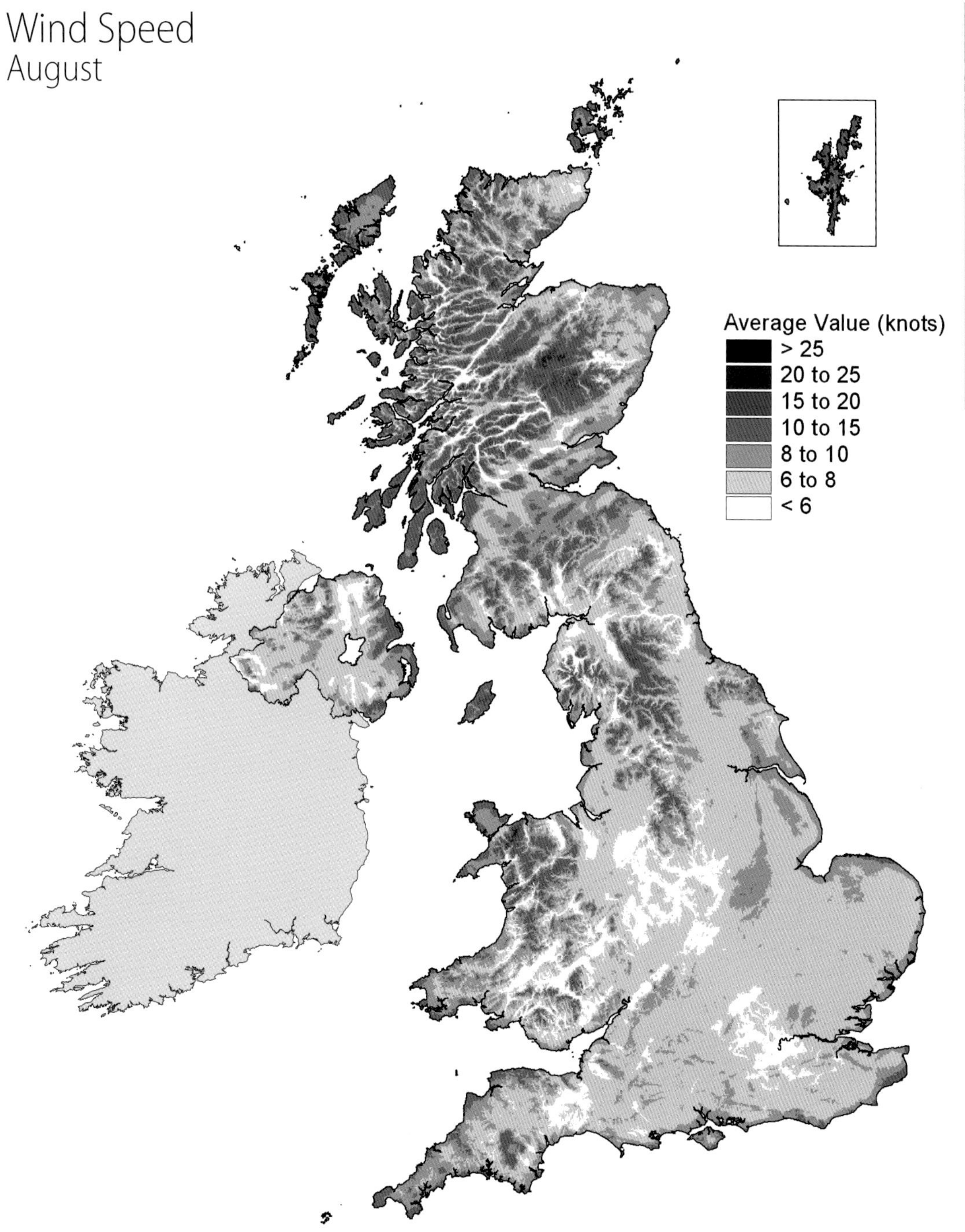

Average Value (knots)
> 25
20 to 25
15 to 20
10 to 15
8 to 10
6 to 8
< 6

Wind Speed
September

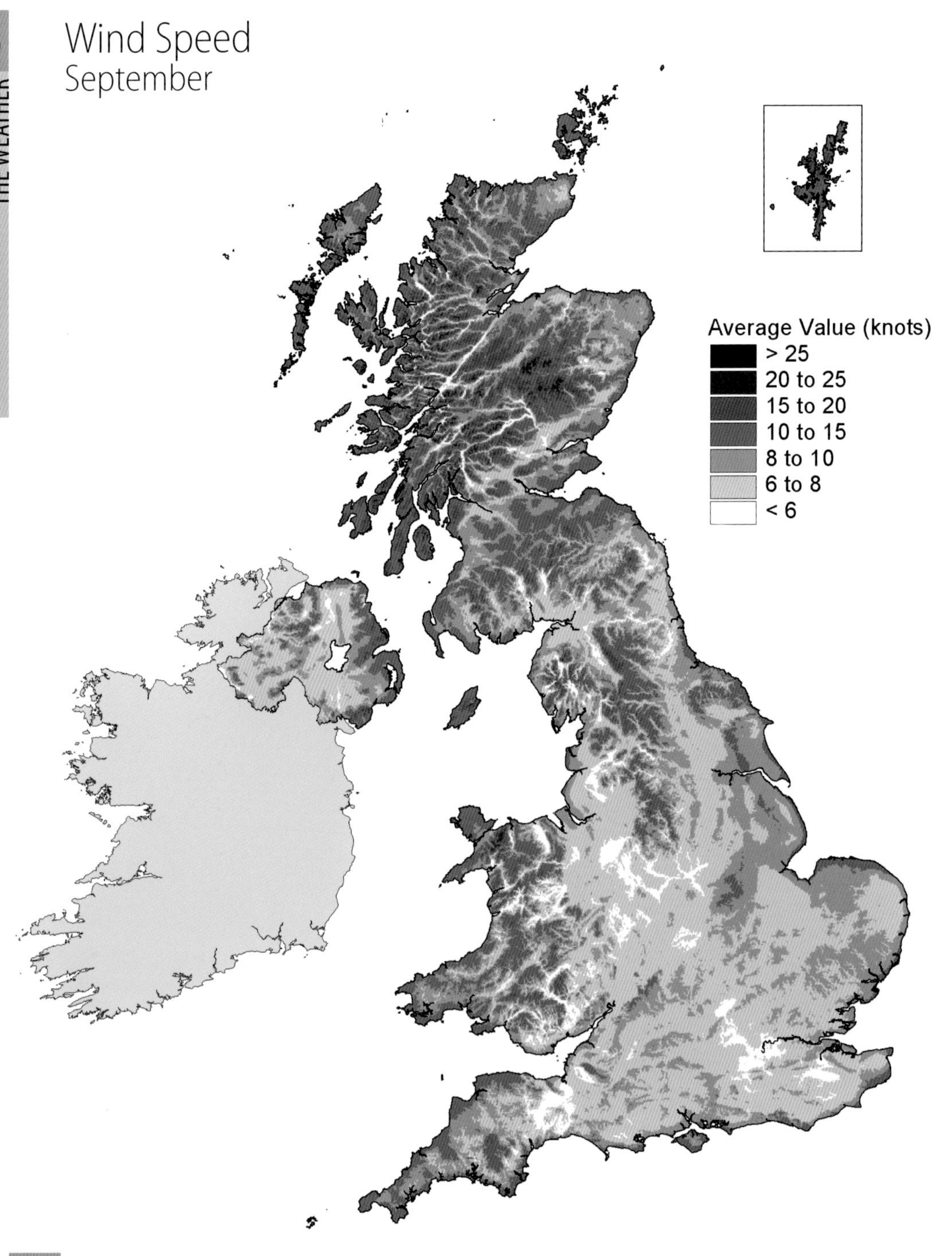
Average Value (knots)
> 25
20 to 25
15 to 20
10 to 15
8 to 10
6 to 8
< 6

Wind Speed
October

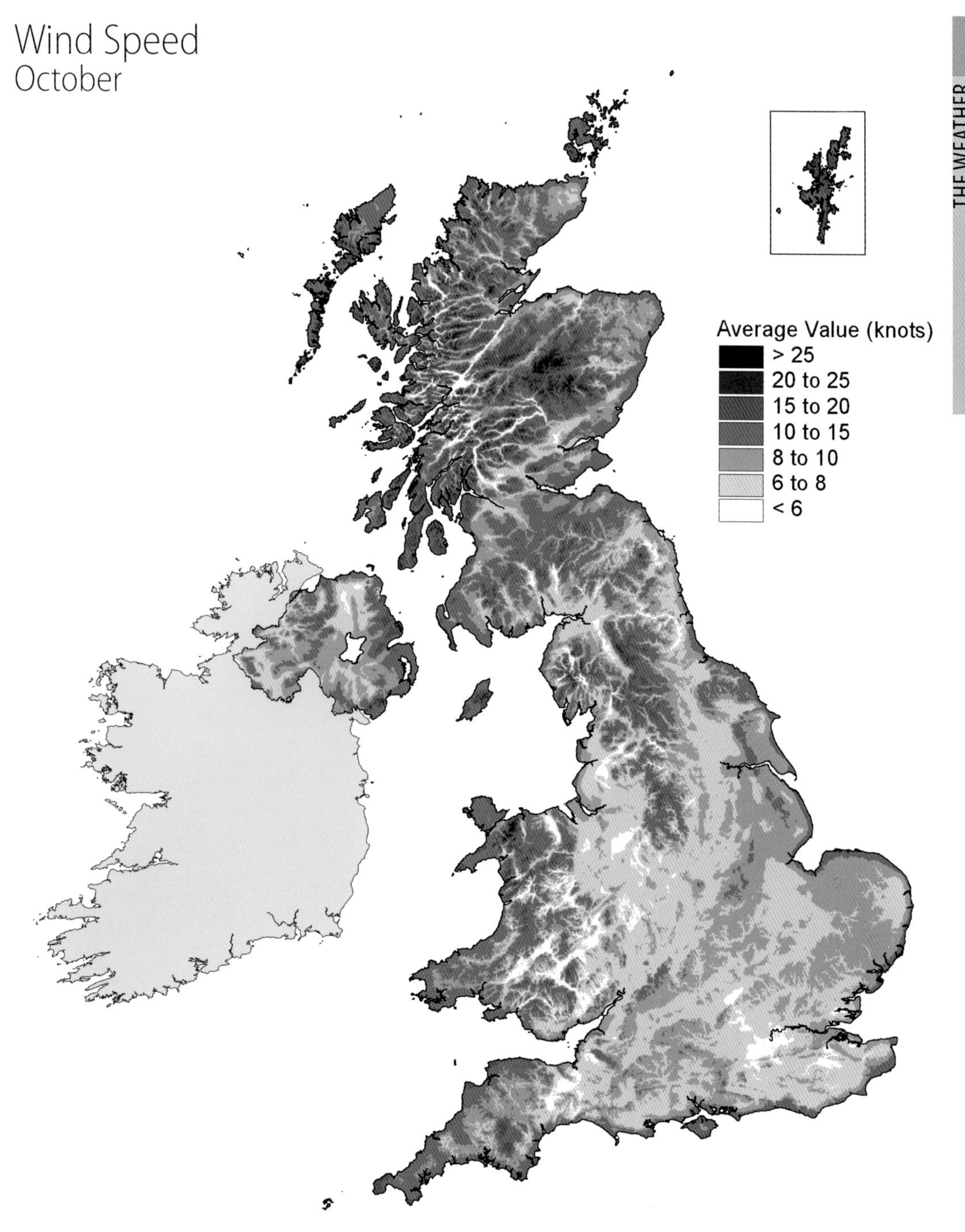
Average Value (knots)
> 25
20 to 25
15 to 20
10 to 15
8 to 10
6 to 8
< 6

Wind Speed
November

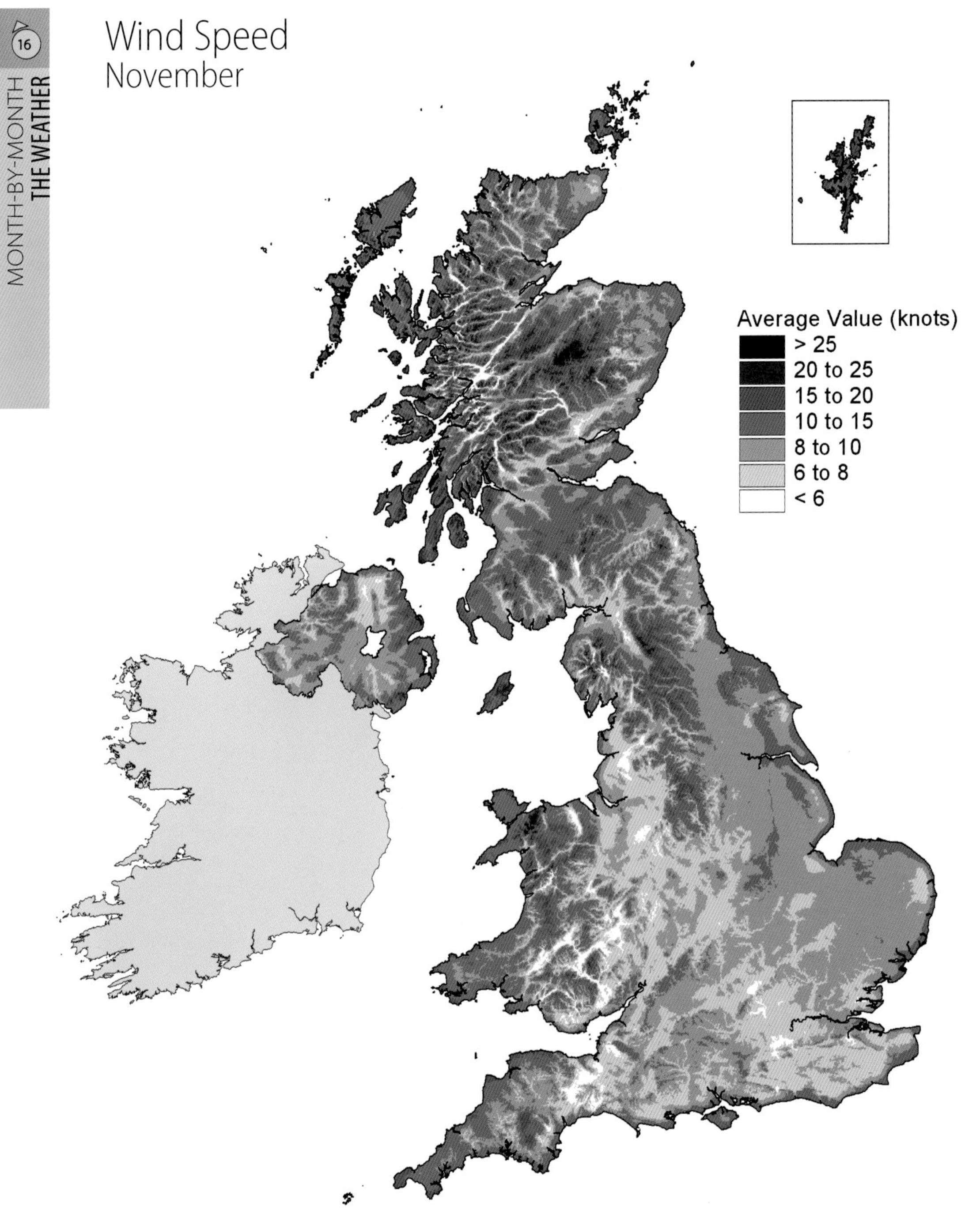
Average Value (knots)
> 25
20 to 25
15 to 20
10 to 15
8 to 10
6 to 8
< 6

Wind Speed
December

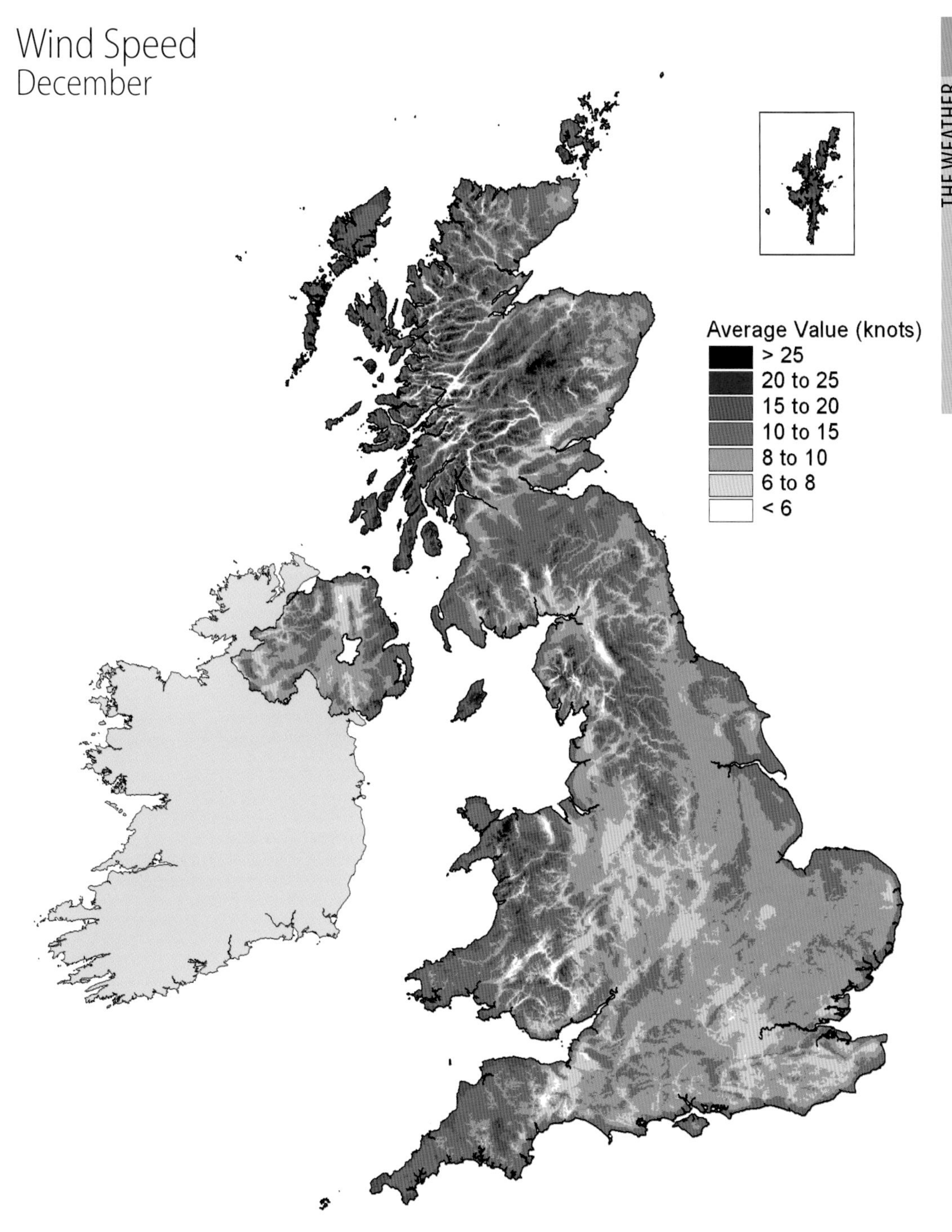
Average Value (knots)
> 25
20 to 25
15 to 20
10 to 15
8 to 10
6 to 8
< 6

Sunshine and Temperature by Season

'The sun has got his hat on, hip-hip-hip-hooray. The sun has got his hat on and he's coming out today'.
Me and My Girl, musical,1937

The trends in sunshine and temperature through the seasons usually go hand-in-hand, but there are some exceptions. Winter is the coldest and dullest season and whilst the summer months are the warmest they are not the sunniest everywhere – late spring is favoured along the western seaboard, especially in Scotland.

Proximity to the Atlantic Ocean and to mainland Europe is the key factor affecting temperature. In winter, western areas are usually mild, thanks to the warm waters of the Atlantic. Any prolonged cold weather is normally linked to cold air arriving from Europe, so eastern areas of Britain are more prone to chilly weather. In summer, this is reversed – the hottest weather affects southern and eastern areas while it's cooler in the west and north.

Location adds another dimension, as altitude, the density of buildings and the distance from the coast all modify a region's weather pattern. Whatever the season, temperatures tend to go down as altitude goes up and urban areas tend to be warmer than their surroundings. However, the fortunes of coastal sites vary – compared to places well inland, they are usually warmer in winter but cooler in summer.

These different effects show up especially clearly in surface temperature measurements. Those taken on grass stay relatively high on coasts and in cities all year round but in rural 'frost hollows' well inland, low values can occur even in spring and summer.

Hours of Sunshine
by Season

These maps confirm that places in the south and near the coast are where you are most likely to get a suntan. Best of all for hours of sunshine are places beside the English Channel, in summer. Over England and Wales the sunniest summer of modern times was 1976 whereas in Scotland and Northern Ireland it was 1995, each with over a third more hours of sunshine than normal. The Clean Air Acts and a decline in heavy industry have led to sunnier winters in our cities.

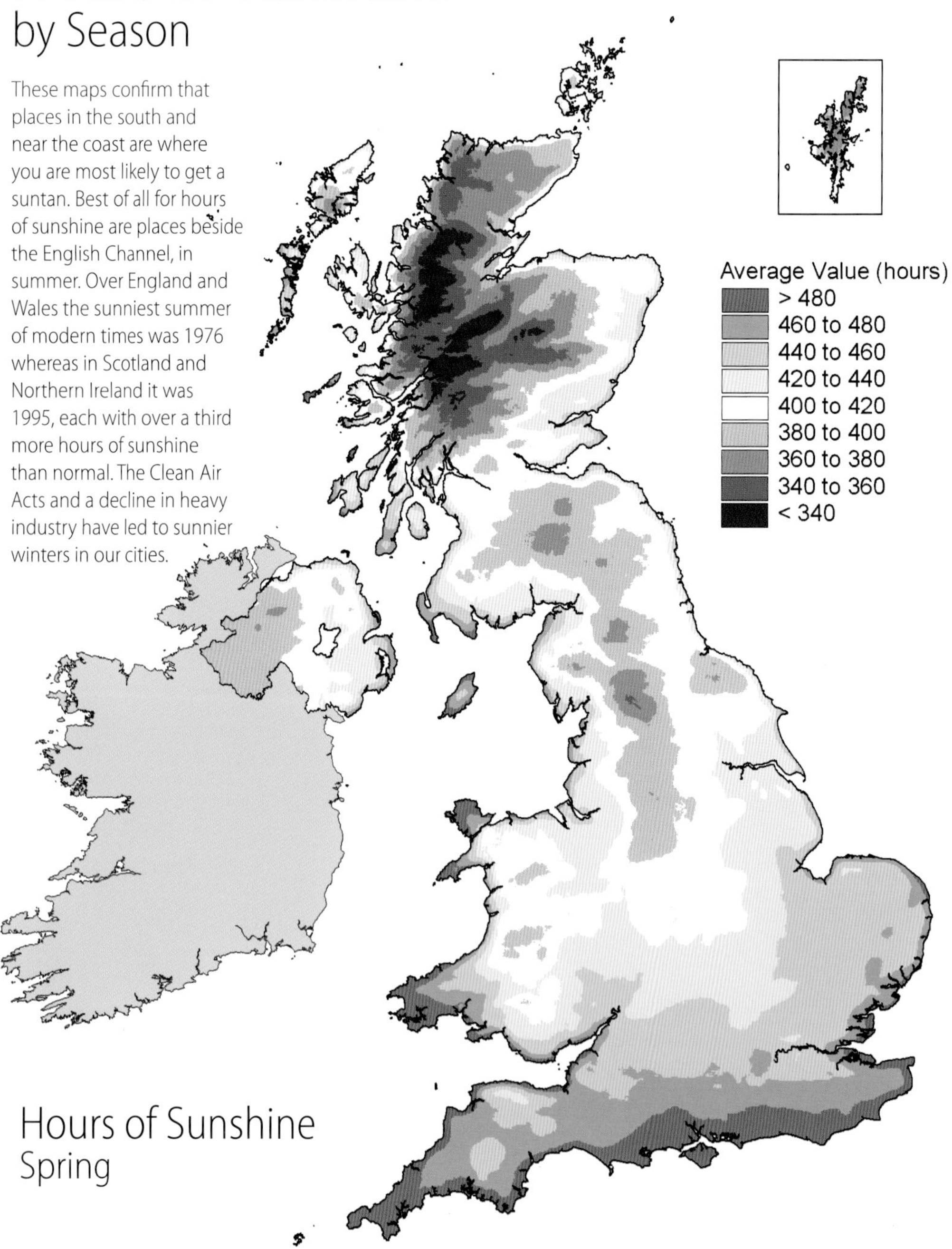
Average Value (hours)
> 480
460 to 480
440 to 460
420 to 440
400 to 420
380 to 400
360 to 380
340 to 360
< 340

Hours of Sunshine
Spring

Hours of Sunshine
Summer

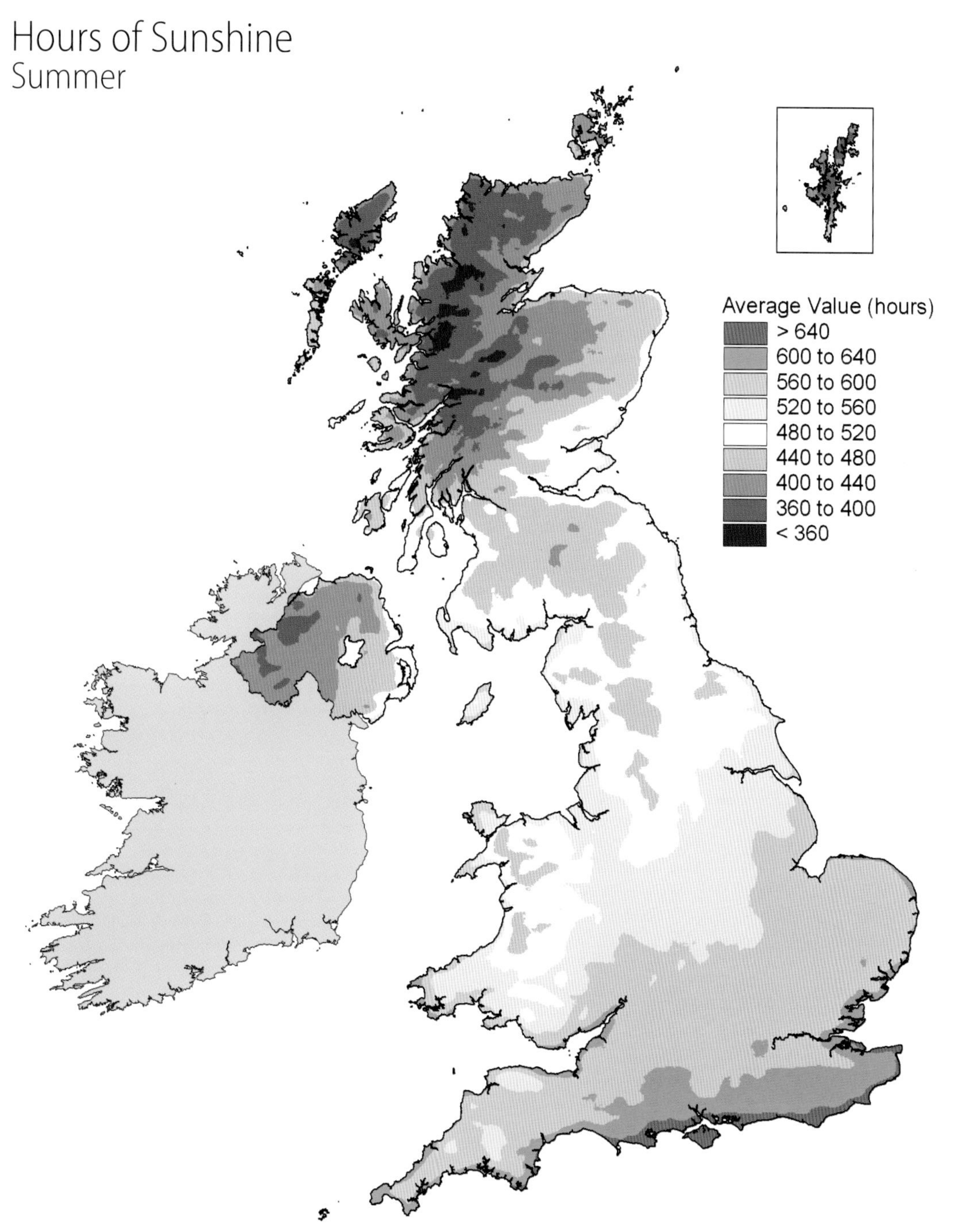

Average Value (hours)
> 640
600 to 640
560 to 600
520 to 560
480 to 520
440 to 480
400 to 440
360 to 400
< 360

Hours of Sunshine
Autumn

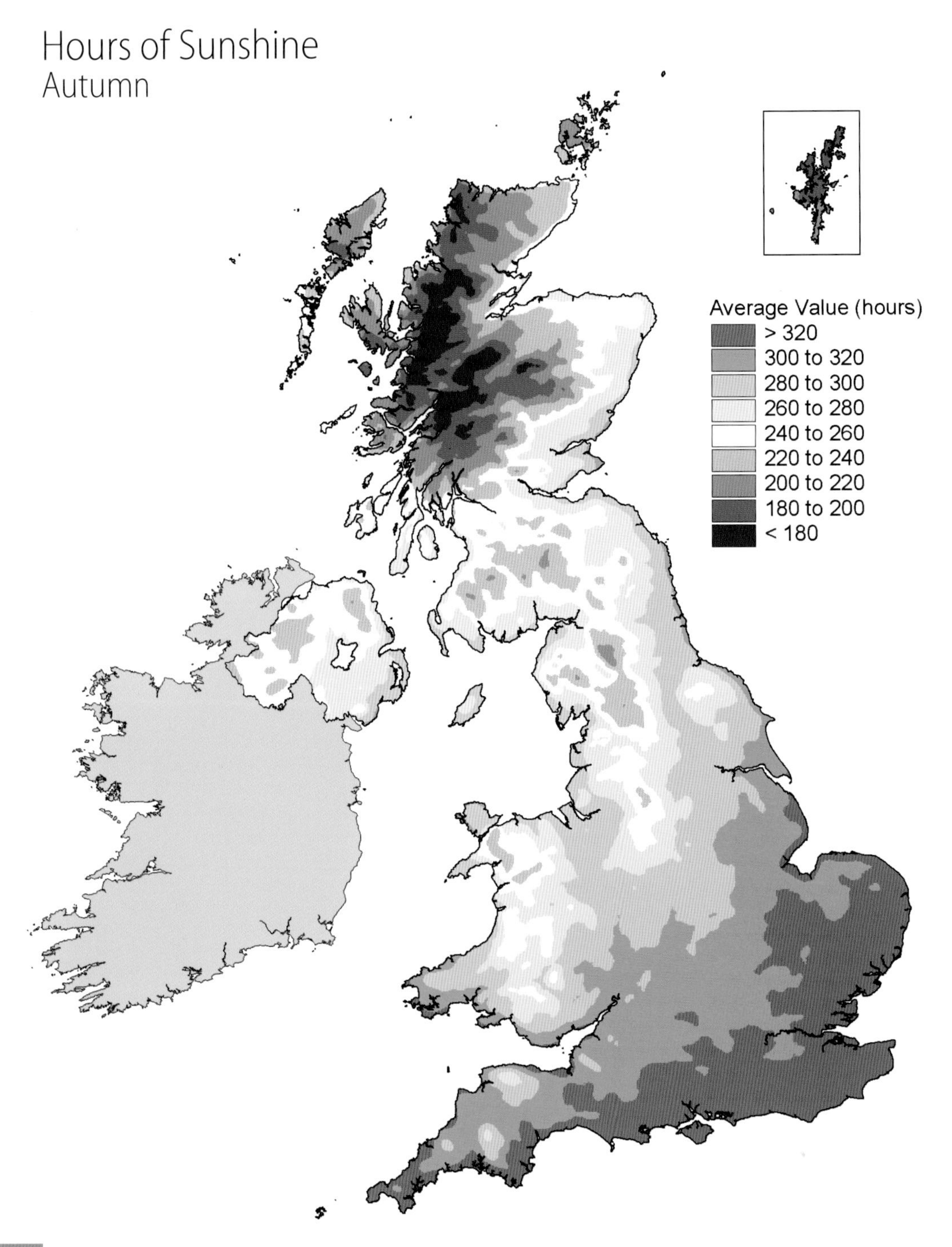
Average Value (hours)
> 320
300 to 320
280 to 300
260 to 280
240 to 260
220 to 240
200 to 220
180 to 200
< 180

Hours of Sunshine
Winter

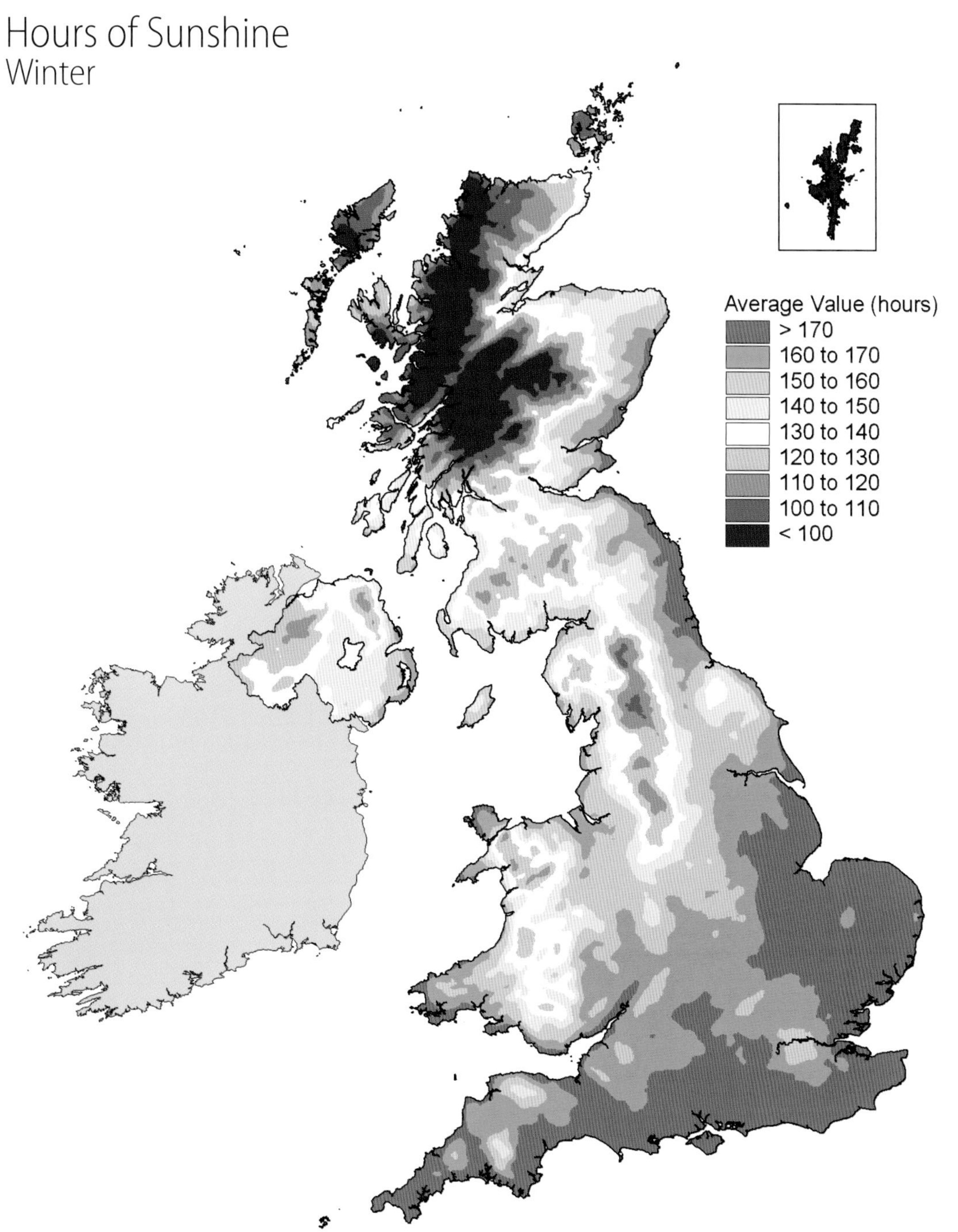

Average Value (hours)
> 170
160 to 170
150 to 160
140 to 150
130 to 140
120 to 130
110 to 120
100 to 110
< 100

Mean Temperature
by Season

These seasonal maps show that the coastal and urban areas are consistently the warmest places, with the south and west favoured in winter and the south-east in summer. In contrast, the Scottish mainland and upland areas in other parts of the UK are always the coolest. Among the memorably warm summers over the UK are those of 2006, 2003 and 1976. The winters of 1963 and 1947 are remembered for their exceptionally cold weather, with temperatures generally staying below freezing for a prolonged time.

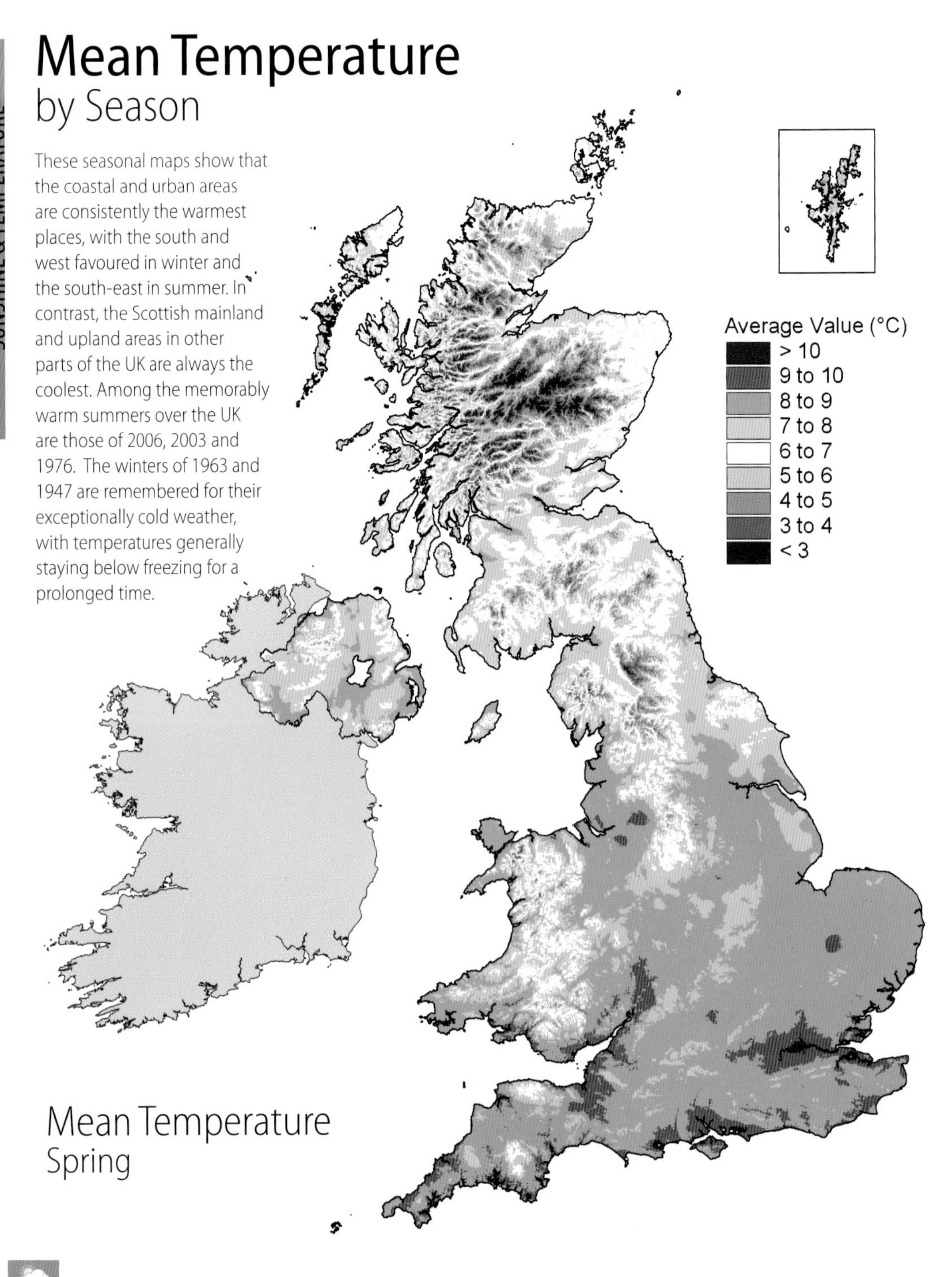
Average Value (°C)
> 10
9 to 10
8 to 9
7 to 8
6 to 7
5 to 6
4 to 5
3 to 4
< 3

Mean Temperature
Spring

Mean Temperature
Summer

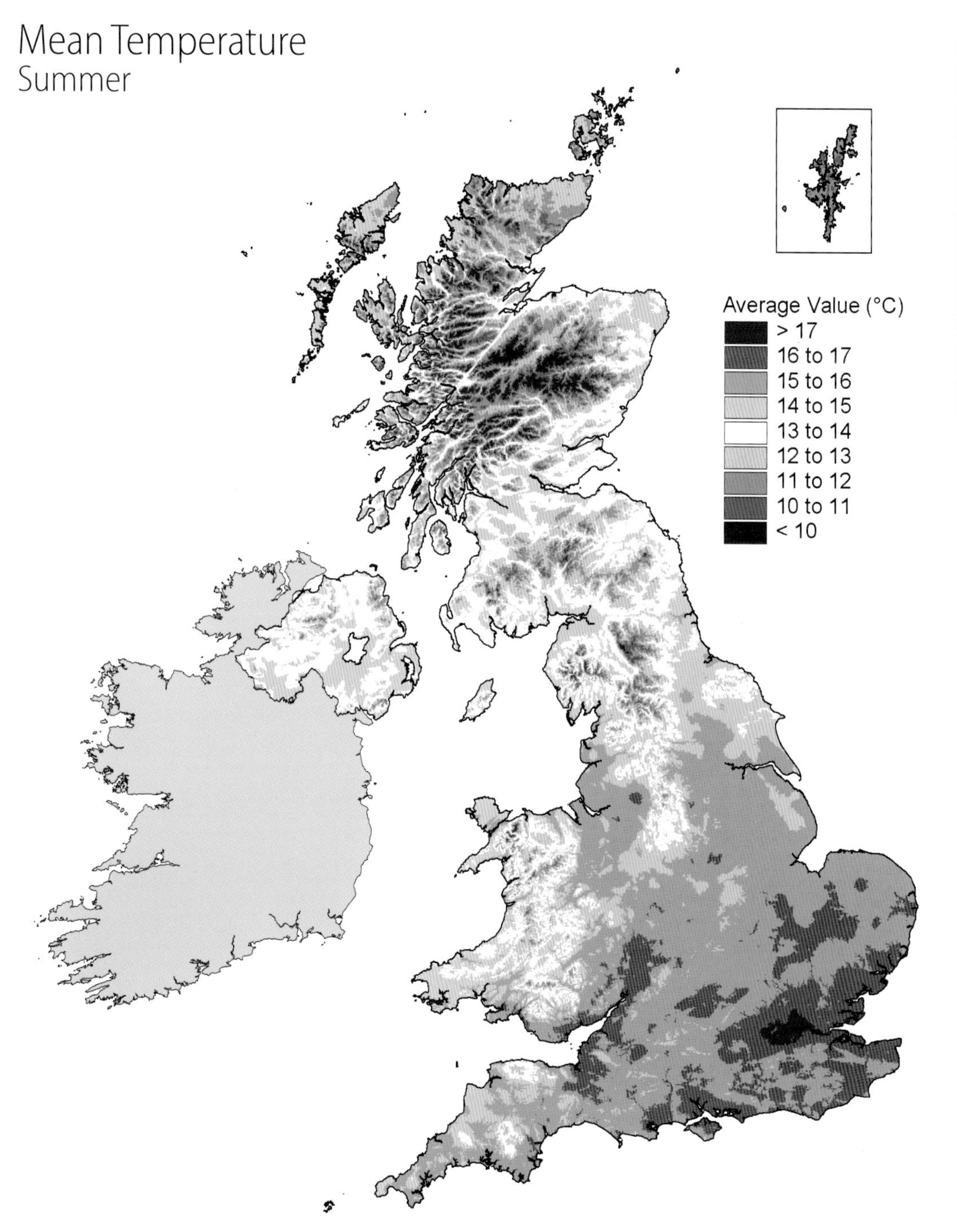

Average Value (°C)
> 17
16 to 17
15 to 16
14 to 15
13 to 14
12 to 13
11 to 12
10 to 11
< 10

Mean Temperature
Autumn

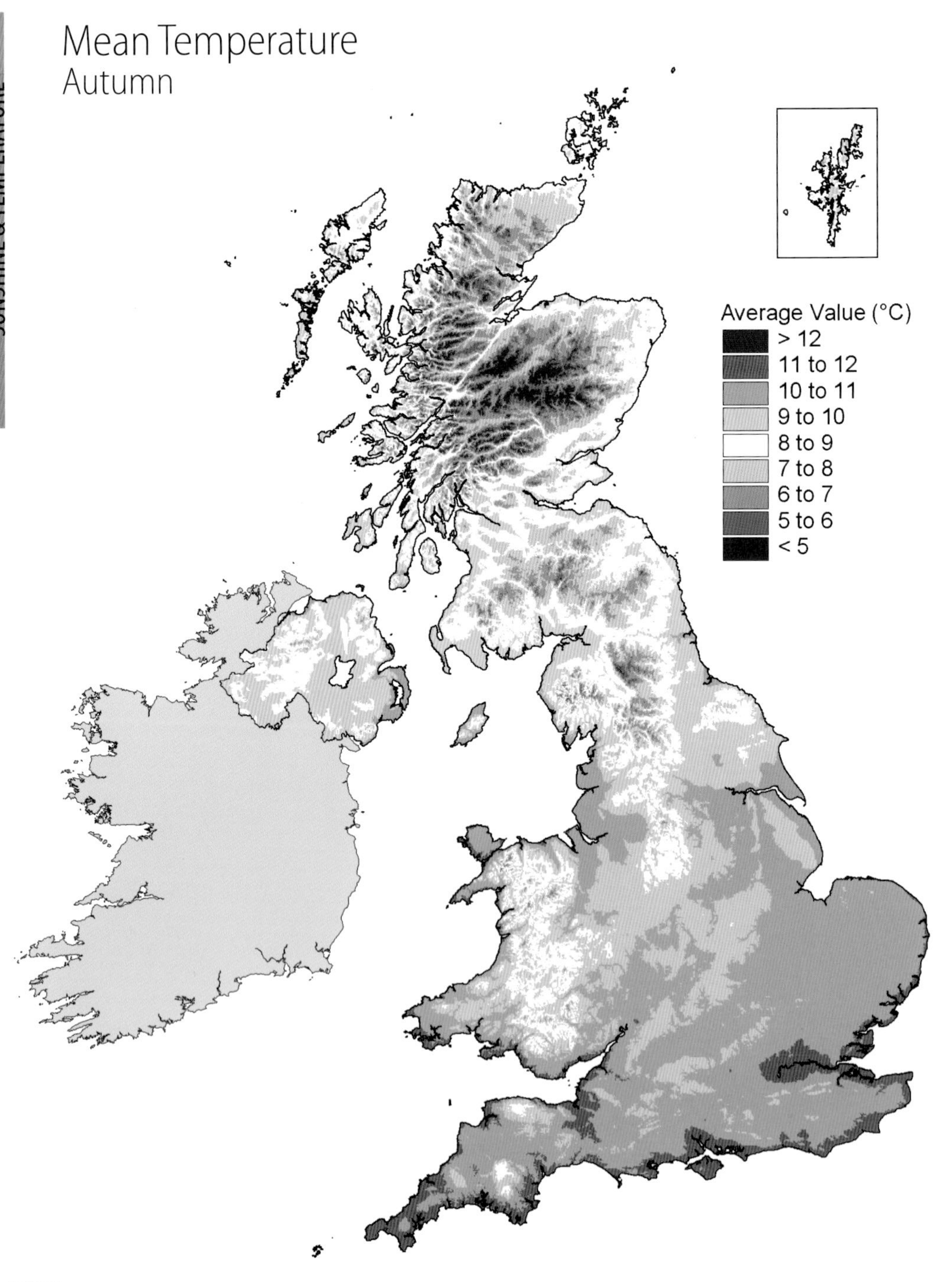
Average Value (°C)
> 12
11 to 12
10 to 11
9 to 10
8 to 9
7 to 8
6 to 7
5 to 6
< 5

Mean Temperature
Winter

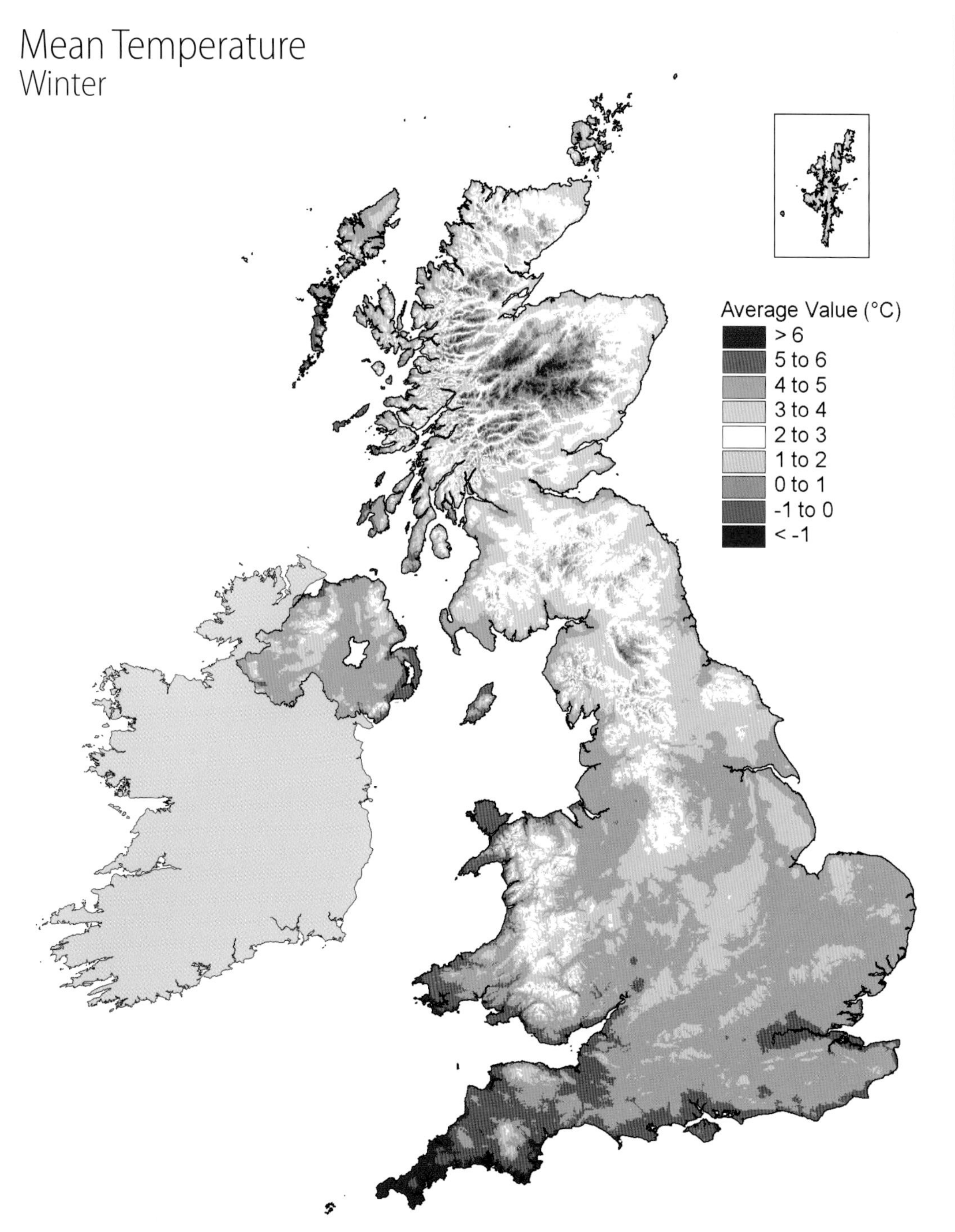

Average Value (°C)
> 6
5 to 6
4 to 5
3 to 4
2 to 3
1 to 2
0 to 1
-1 to 0
< -1

Grass Minimum Temperature
by Season

In all seasons, cities and coasts show up as being the least prone to low surface-temperatures. In summer, cooler conditions are confined to upland areas but ground frosts can occur occasionally in frost hollows elsewhere. In winter, all except a few coastal areas have mean values below freezing. Ground frosts are common across the UK, typically occurring on about half of all nights in winter in the far south of England and on around two-thirds of winter nights across Scotland.

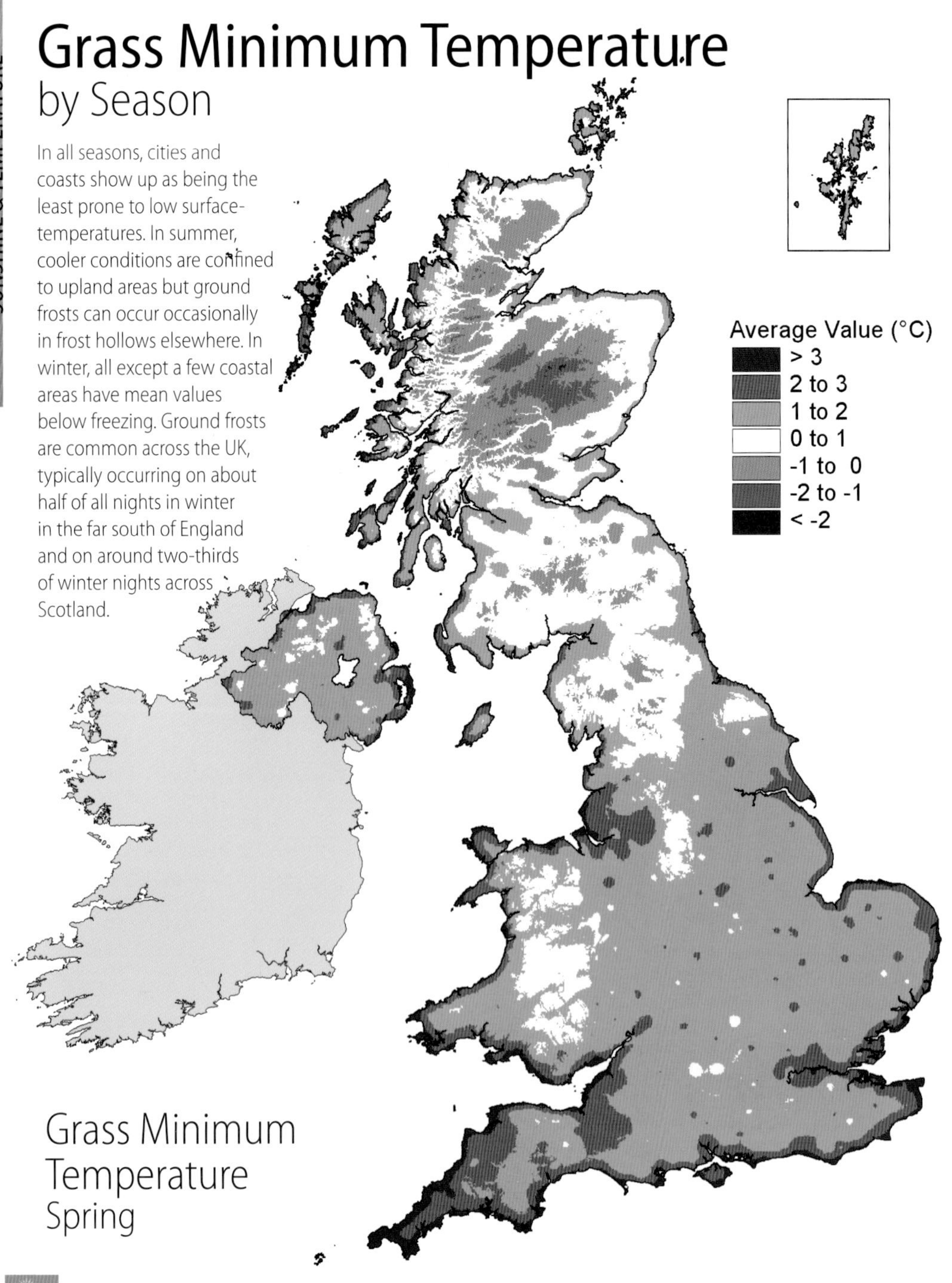
Average Value (°C)
> 3
2 to 3
1 to 2
0 to 1
-1 to 0
-2 to -1
< -2

Grass Minimum Temperature
Spring

Grass Minimum Temperature
Summer

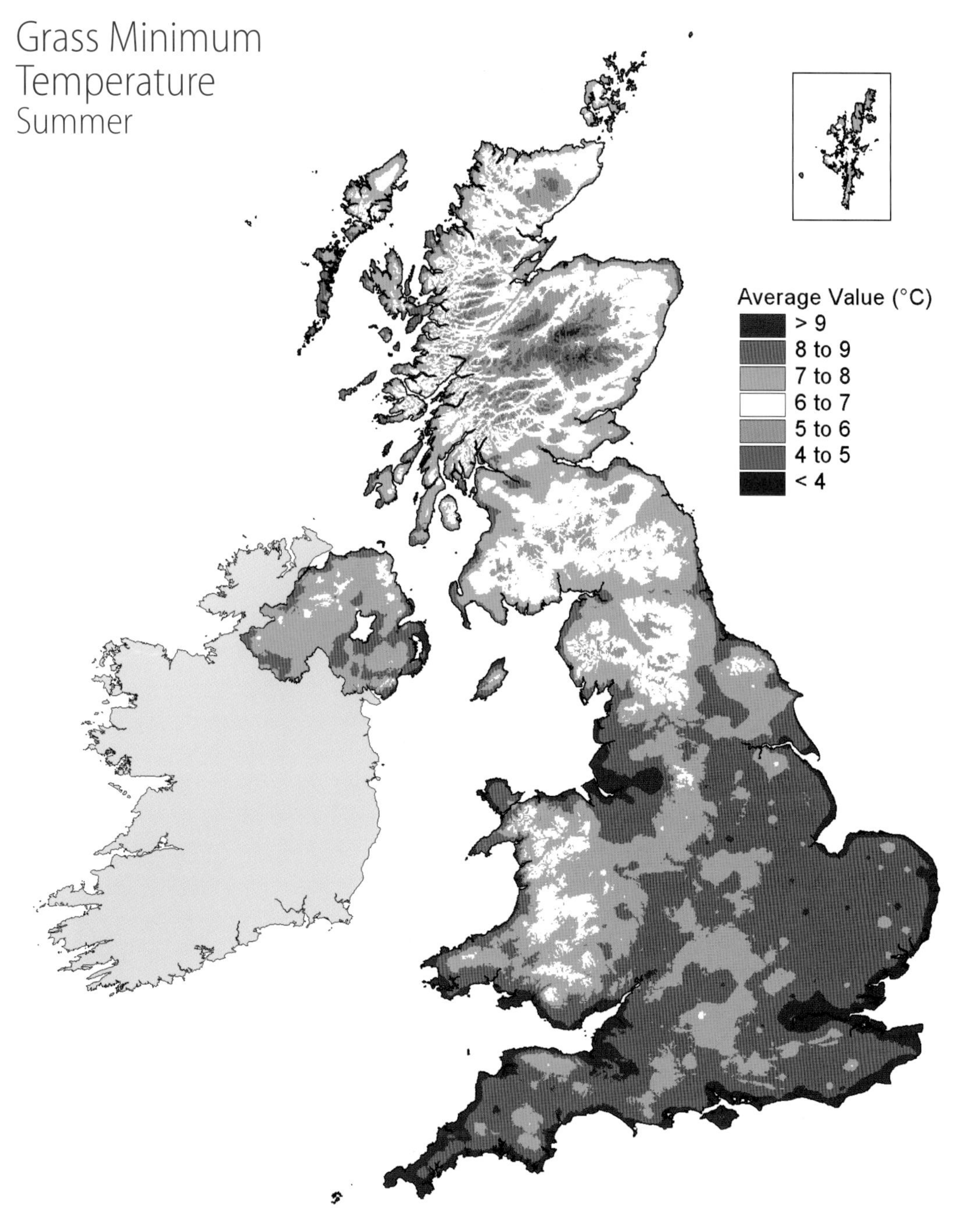
Average Value (°C)
> 9
8 to 9
7 to 8
6 to 7
5 to 6
4 to 5
< 4

Grass Minimum Temperature
Autumn

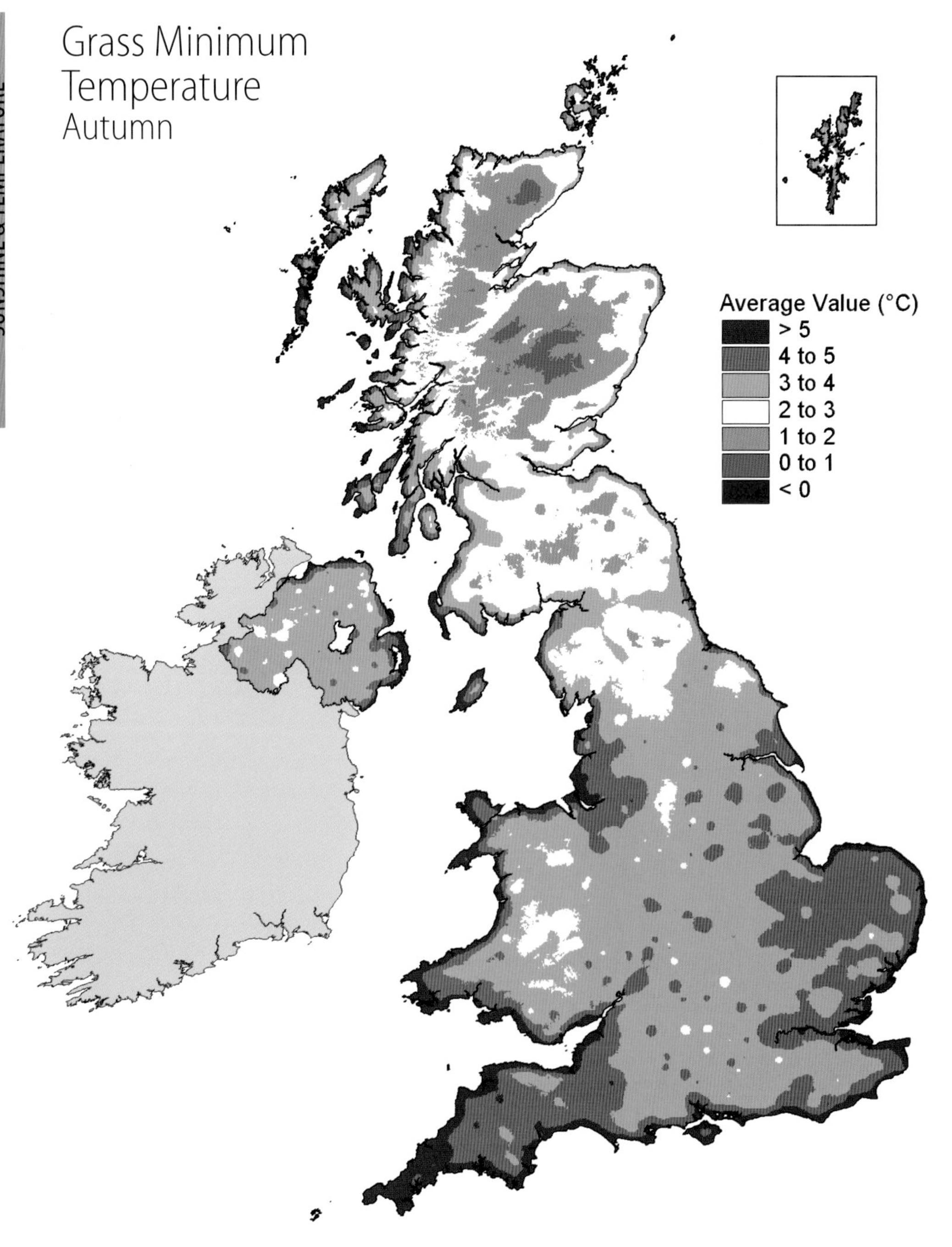
Average Value (°C)
> 5
4 to 5
3 to 4
2 to 3
1 to 2
0 to 1
< 0

Grass Minimum Temperature
Winter

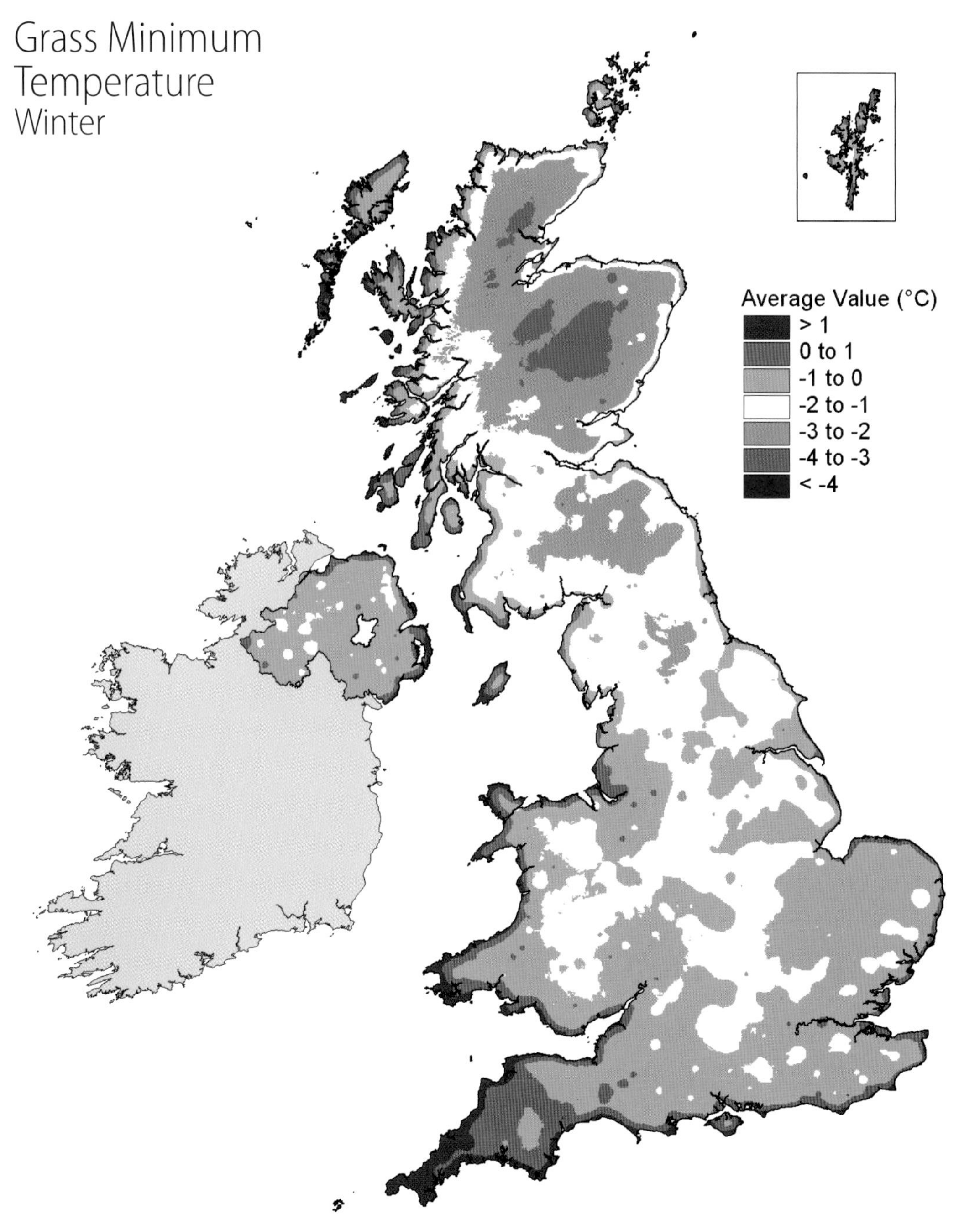
Average Value (°C)
> 1
0 to 1
-1 to 0
-2 to -1
-3 to -2
-4 to -3
< -4

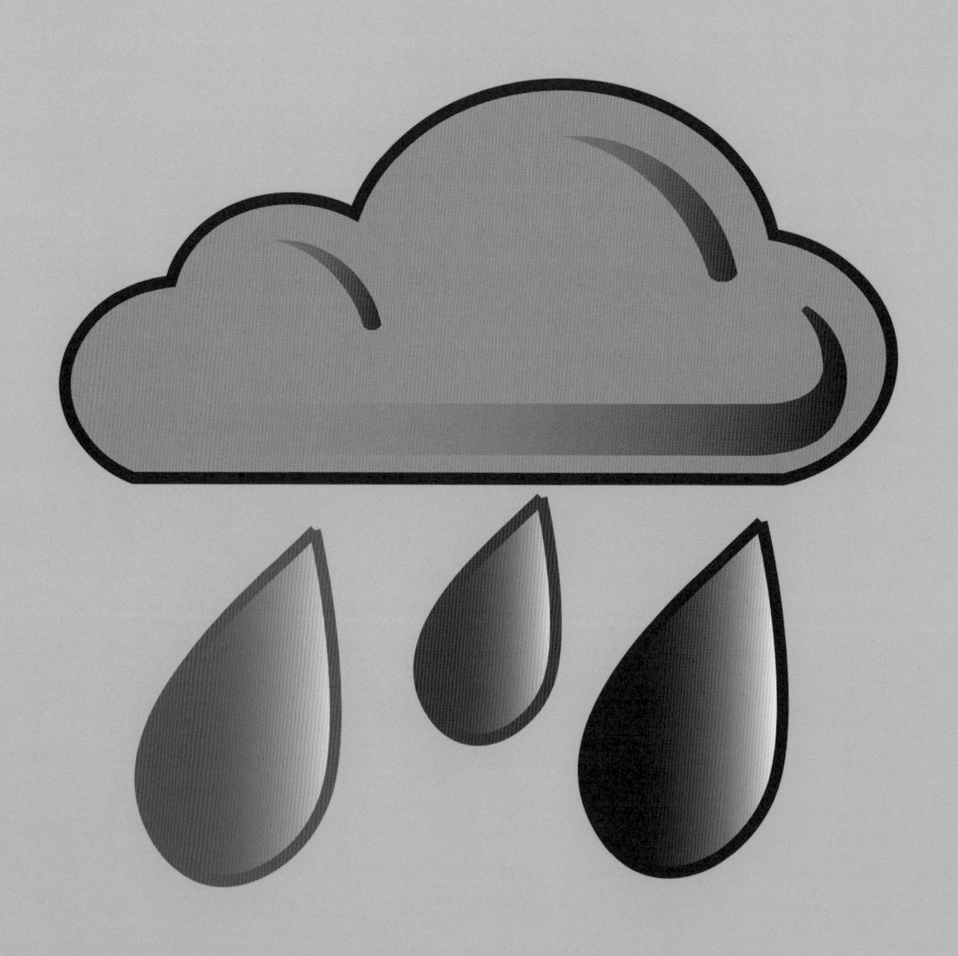

Wind and Rain by Season

'March winds and April showers bring forth May flowers'.
Anon.

The strongest winds usually occur during late autumn and winter, with calmer conditions in spring and summer. March is a transition month, hence its reputation for 'coming in like a lion and going out like a lamb'. Whatever the season, upland areas and coasts are the windiest places, especially those nearest the Atlantic Ocean. These areas also experience the most gales – when mean winds are 34 knots (39 mph) or more. Gales peak in number in winter and are rare in summer.

Western areas usually receive well over half their annual rainfall in the autumn and winter. Towards the east, further from the Atlantic weather systems, the seasonal pattern is more even. Rainfall totals here are also much lower, with annual averages in parts of East Anglia less than a fifth of those in the wettest parts of the north and west. However, there is much less contrast across the country in the number of rainy days – barely a factor of two between the lowest and highest.

The nature of the rain varies throughout the seasons, with weather systems bringing prolonged rain in late autumn and winter but the summer having more showers. These can be intense, particularly inland and in the east. April's showery reputation probably comes from it being the time when showers start to become more frequent ahead of summer. Otherwise, spring is the best season for dry weather.

Wind Speed
by Season

Throughout the year, places along the coast and on higher ground are the windiest – especially those exposed to the west. Even on warm summer days, sea breezes can set in bringing fresher conditions to the coast. Autumn and winter see the strongest winds with frequent gales, particularly in the Scottish Highlands and islands. Occasionally, very strong winds strike further south – the storms of 15-16 October 1987 and 25 January 1990 were two of the most destructive ever to hit the UK.

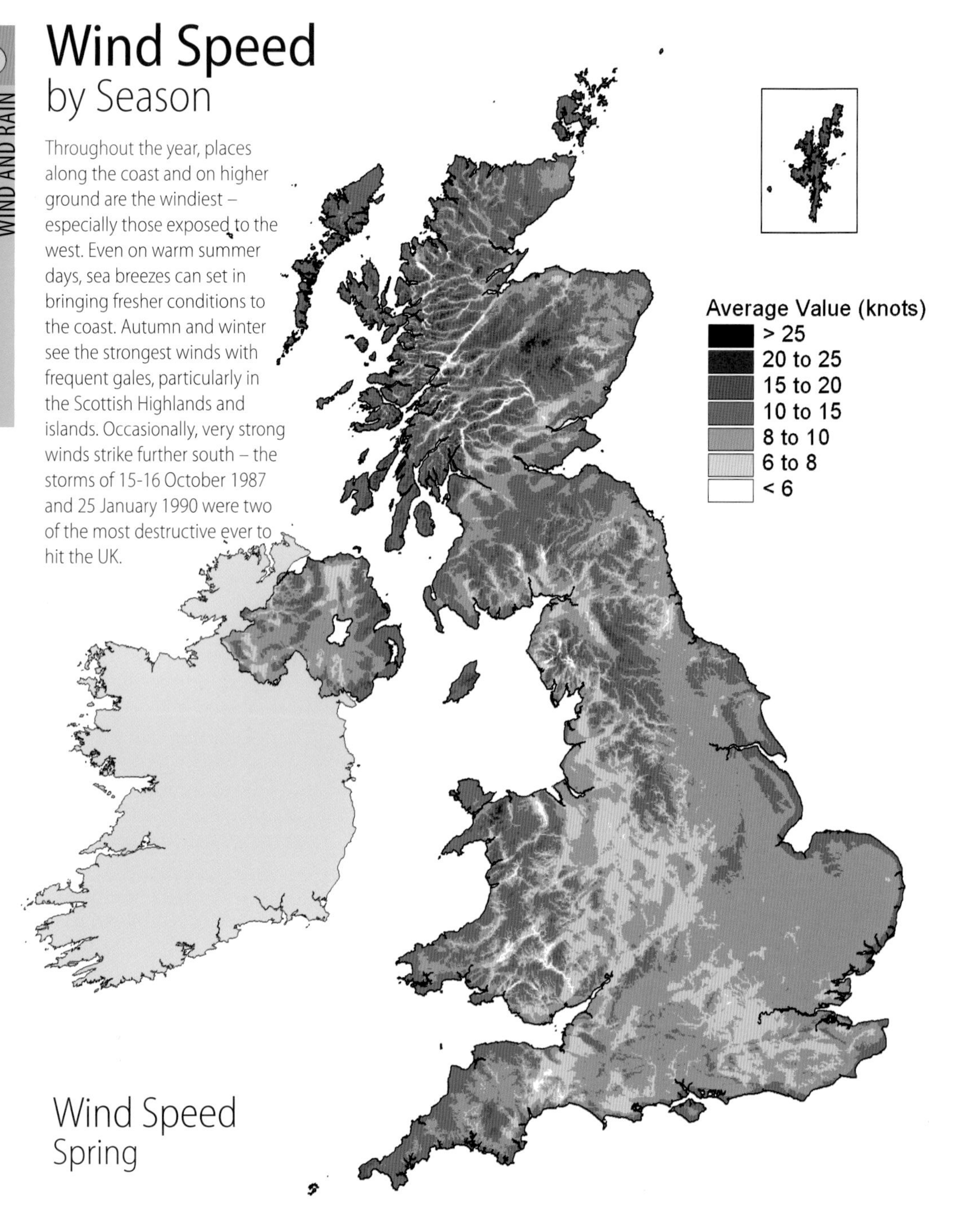
Average Value (knots)
> 25
20 to 25
15 to 20
10 to 15
8 to 10
6 to 8
< 6

Wind Speed
Spring

Wind Speed
Summer

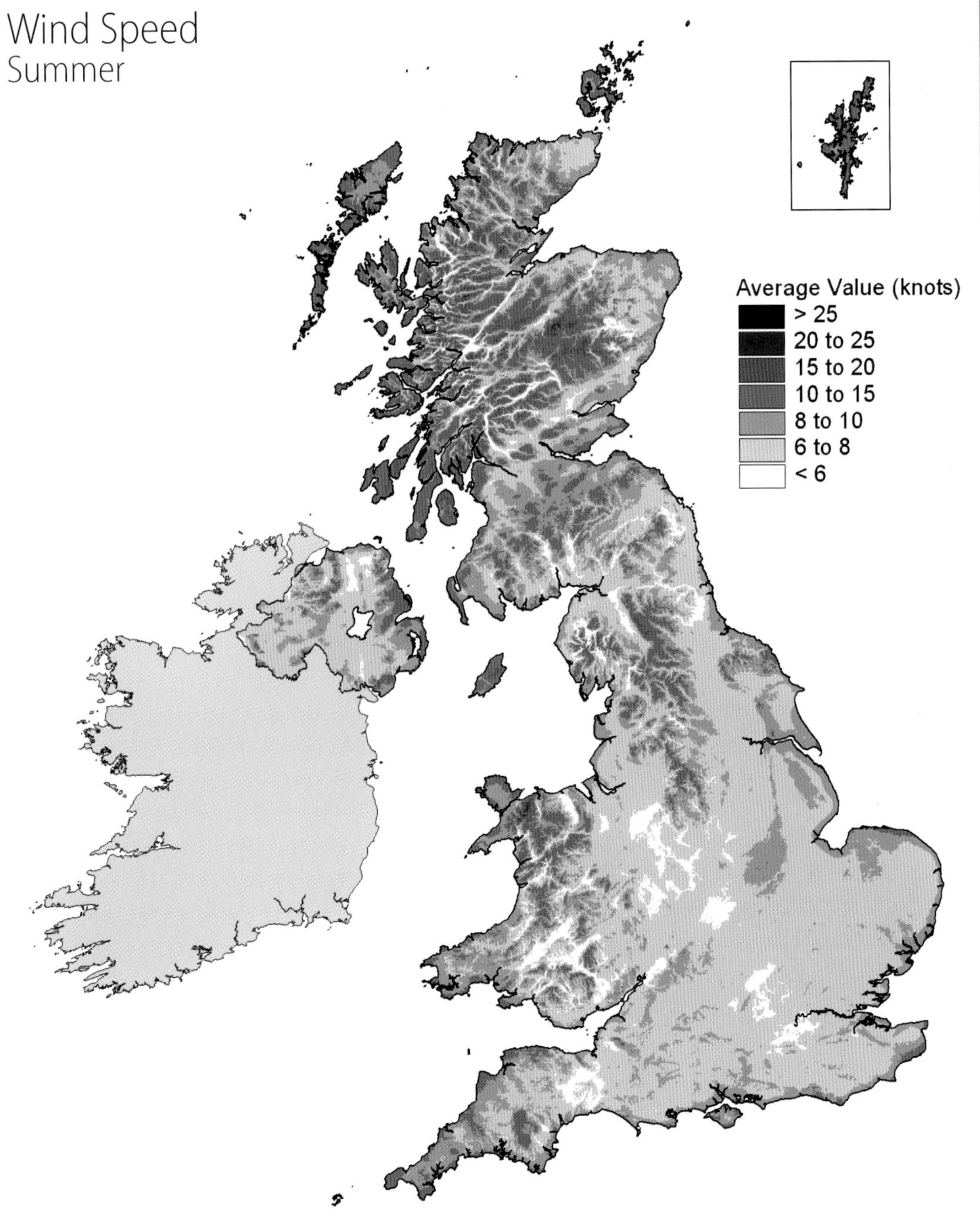
Average Value (knots)
> 25
20 to 25
15 to 20
10 to 15
8 to 10
6 to 8
< 6

Wind Speed
Autumn

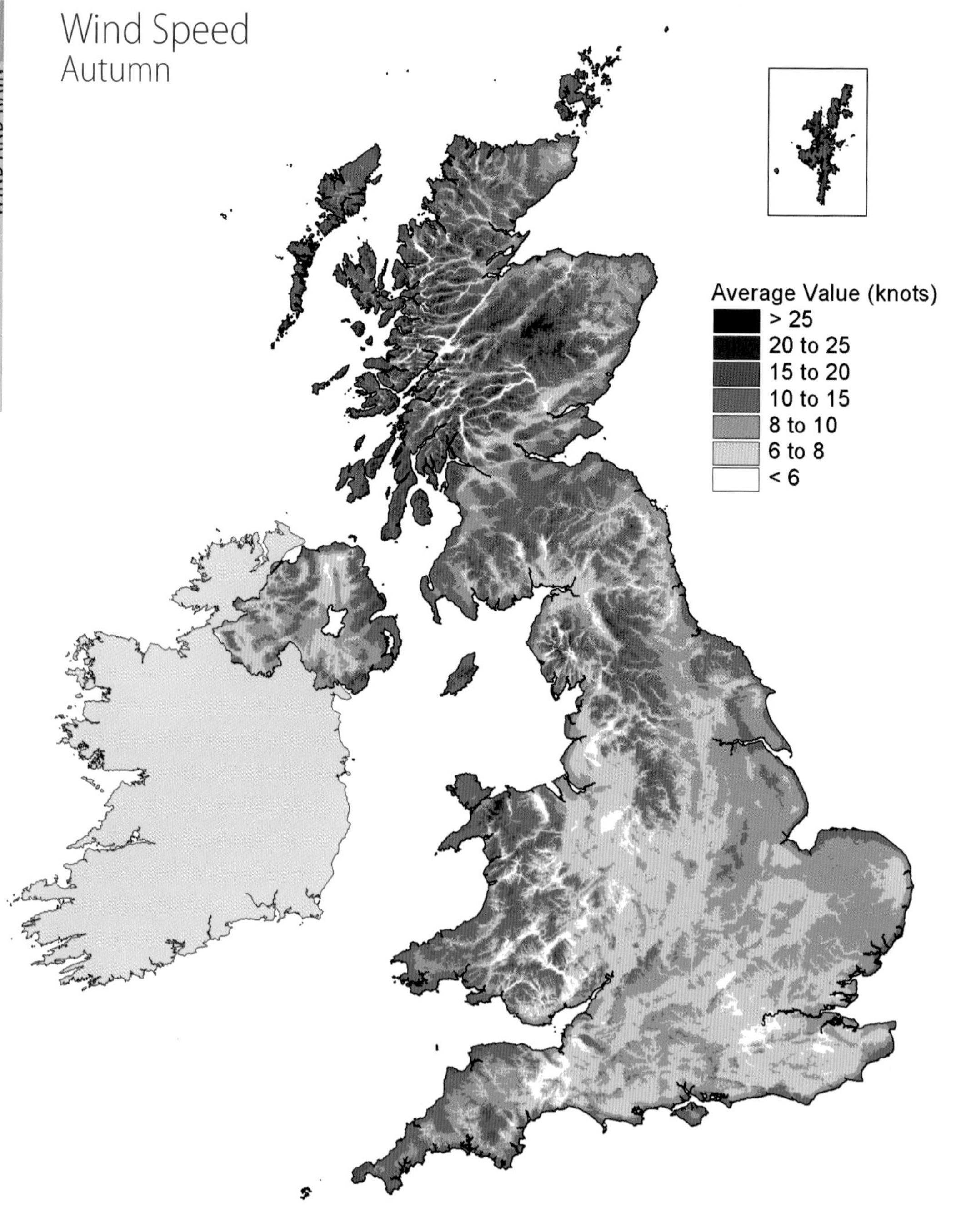
Average Value (knots)
> 25
20 to 25
15 to 20
10 to 15
8 to 10
6 to 8
< 6

Wind Speed
Winter

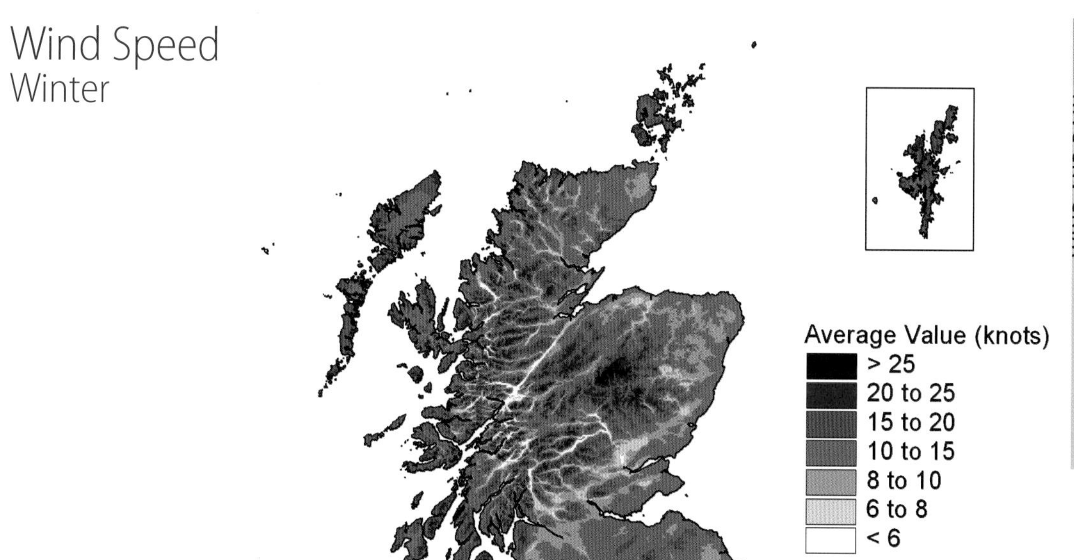
Average Value (knots)
> 25
20 to 25
15 to 20
10 to 15
8 to 10
6 to 8
< 6

Total Rainfall
by Season

These maps reflect the annual cycle of rainfall. In autumn and winter, the hills and mountains of the west and north regularly get a soaking. Everywhere is much drier by spring, particularly in central and eastern parts of the UK. Summer is relatively dry too, but showers can boost the rainfall averages of eastern England. Normally, the flooding risk is greatest in winter, but some summers break this rule – 2007 saw the wettest summer on record across England and Wales with some devastating floods.

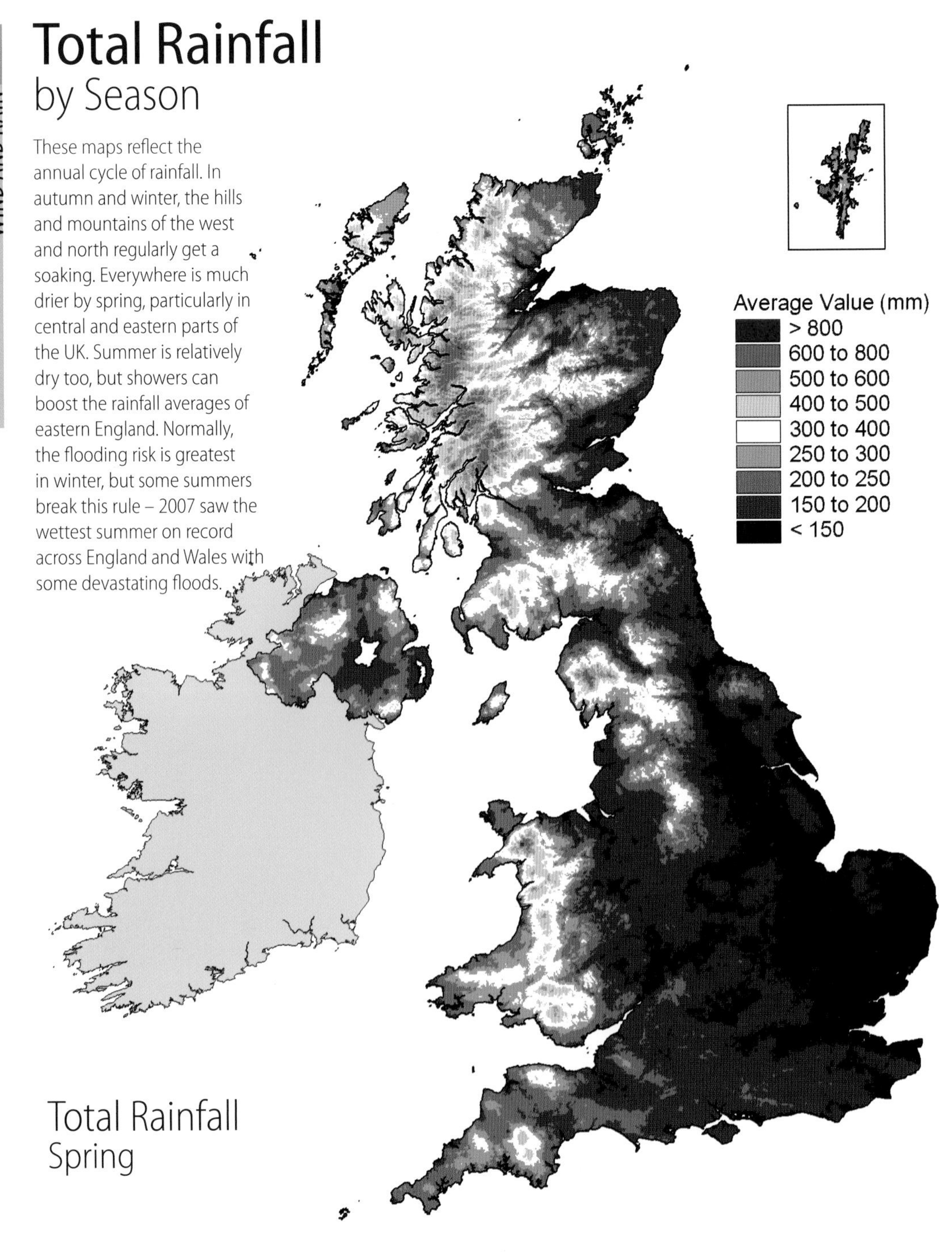
Average Value (mm)
> 800
600 to 800
500 to 600
400 to 500
300 to 400
250 to 300
200 to 250
150 to 200
< 150

Total Rainfall
Spring

Total Rainfall
Summer

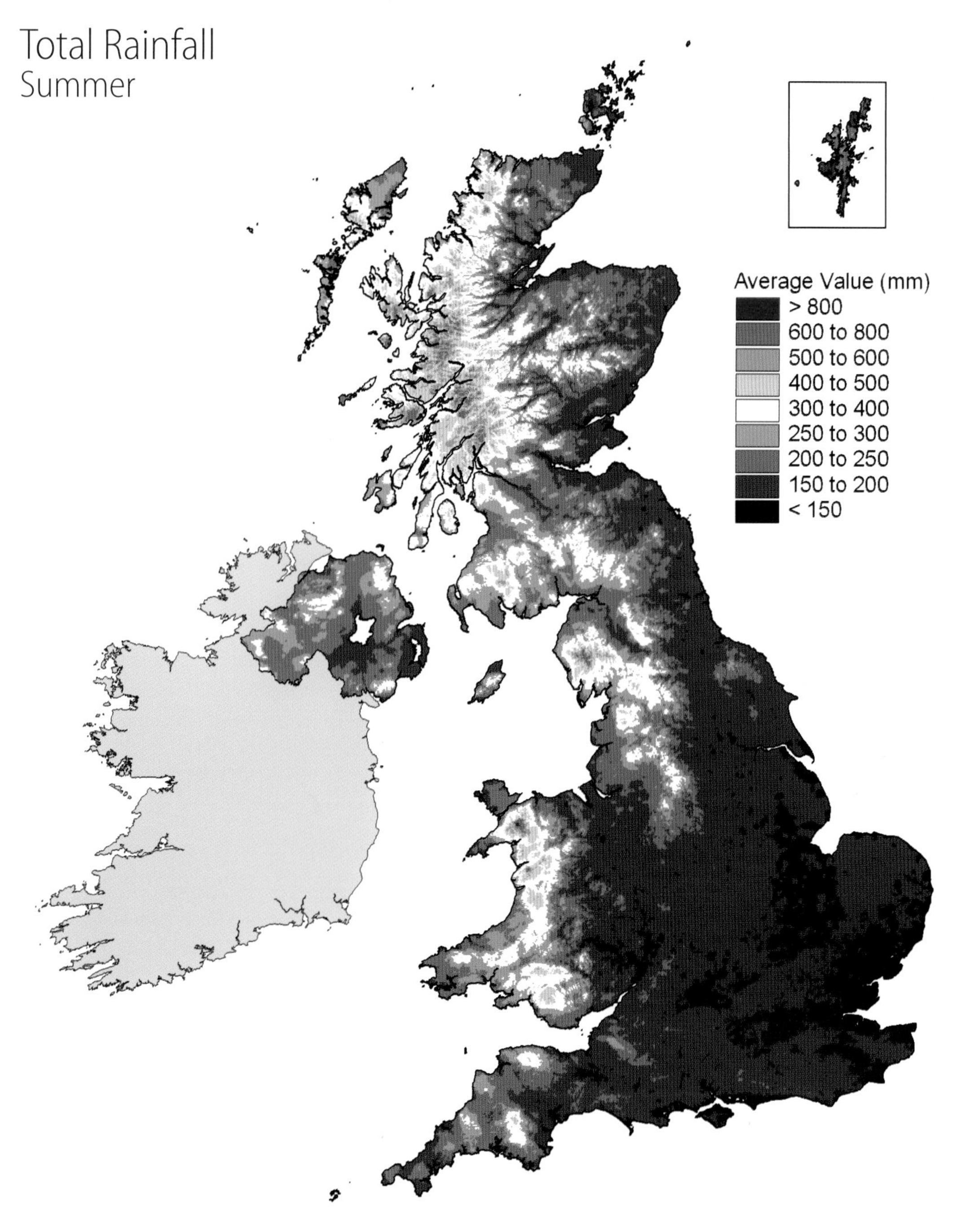
Average Value (mm)
> 800
600 to 800
500 to 600
400 to 500
300 to 400
250 to 300
200 to 250
150 to 200
< 150

Total Rainfall
Autumn

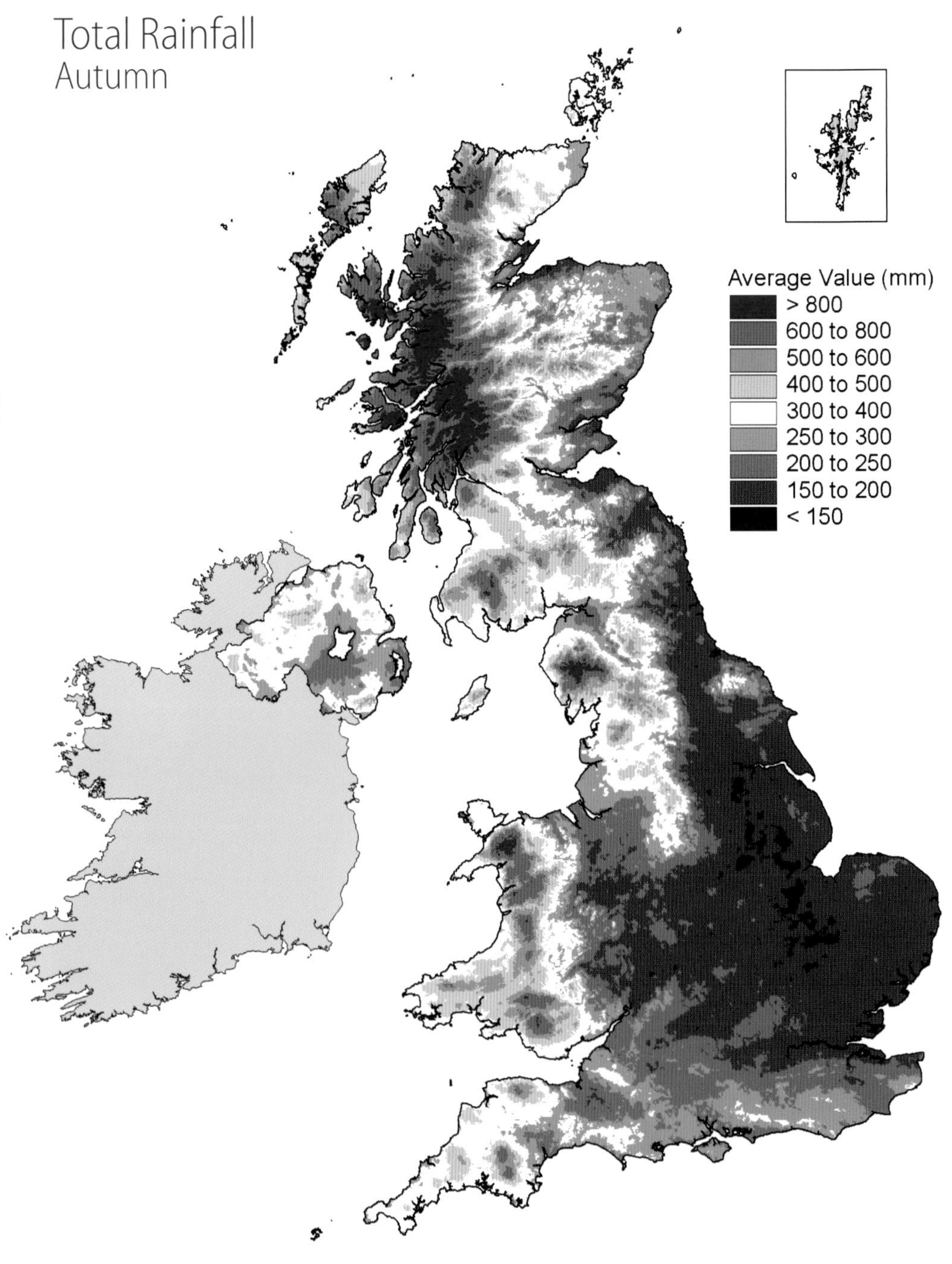
Average Value (mm)
> 800
600 to 800
500 to 600
400 to 500
300 to 400
250 to 300
200 to 250
150 to 200
< 150

Total Rainfall
Winter

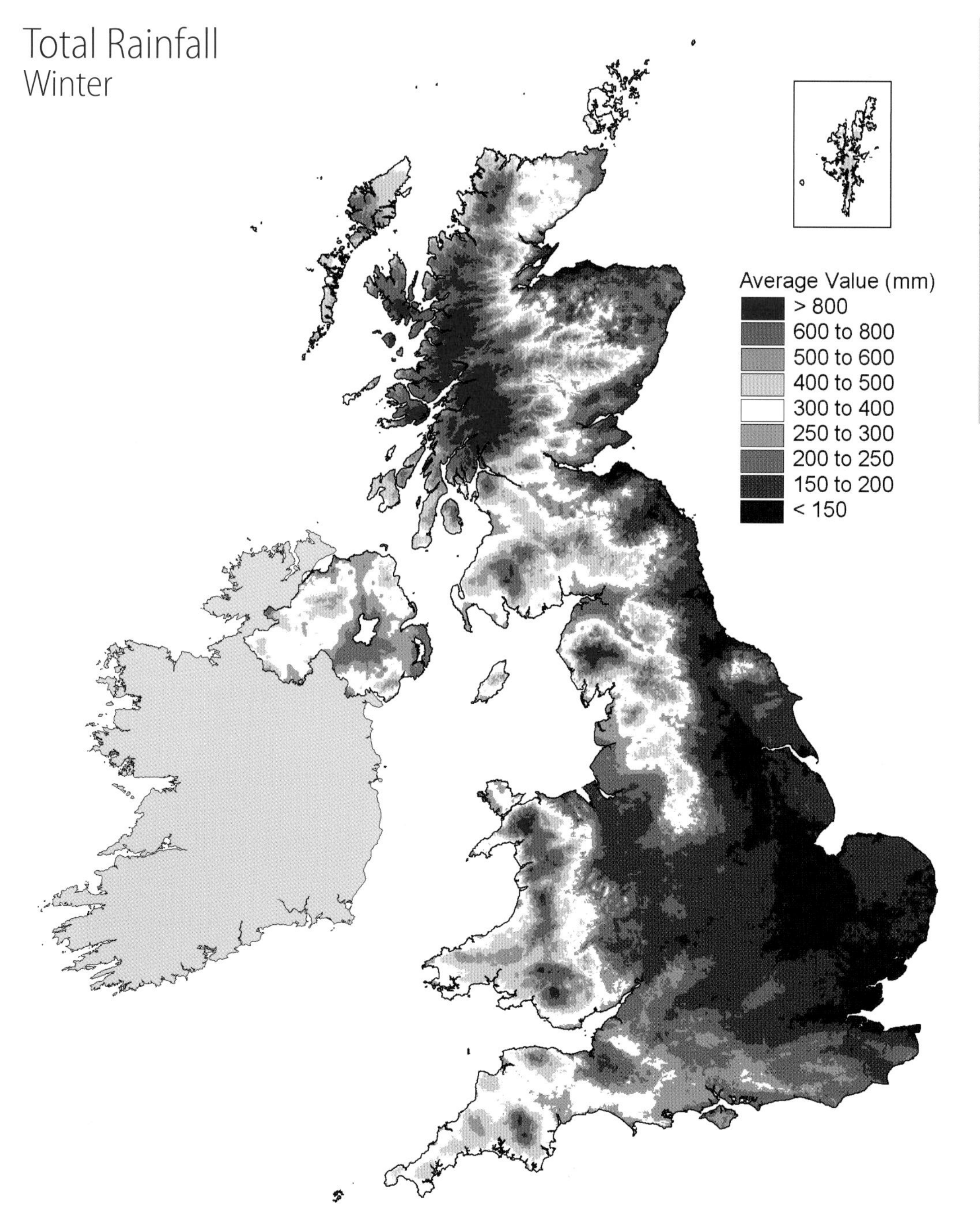

Average Value (mm)
> 800
600 to 800
500 to 600
400 to 500
300 to 400
250 to 300
200 to 250
150 to 200
< 150

Days of Rain 0.2mm or more
by Season

These maps reveal the seasonal pattern of 'rain days', which are those with a light shower or more. Across the south and east of England usually less than one summer day in three has this amount of rainfall whereas in autumn and winter it approaches half the days. In contrast, in autumn and winter much of Scotland, along with upland areas elsewhere, receive rain on at least two days in three.

Average Value (days)

- > 70
- 65 to 70
- 60 to 65
- 55 to 60
- 50 to 55
- 45 to 50
- 40 to 45
- 35 to 40
- < 35

Days of Rain
0.2mm or more
Spring

Days of Rain 0.2mm or more
Summer

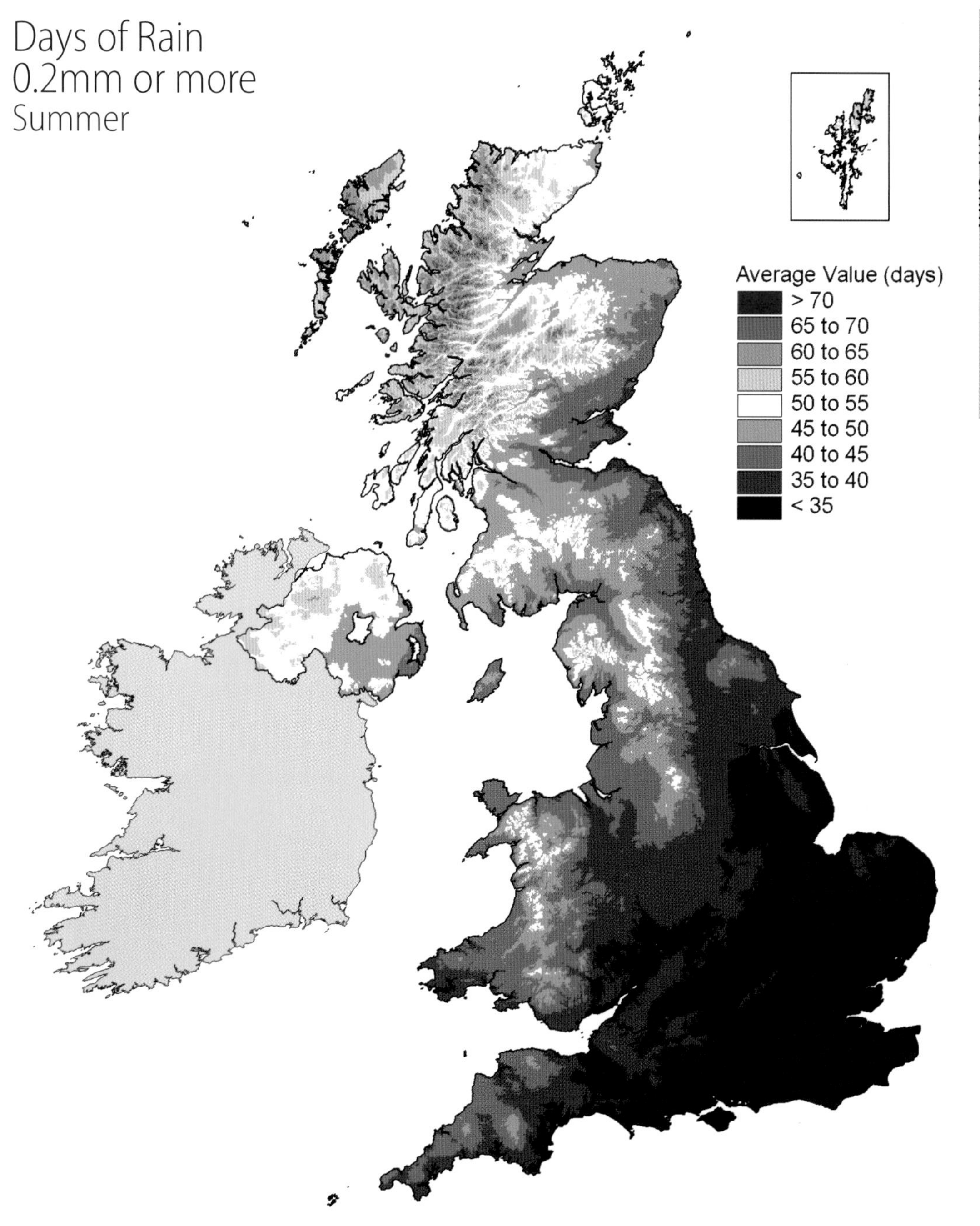
Average Value (days)
> 70
65 to 70
60 to 65
55 to 60
50 to 55
45 to 50
40 to 45
35 to 40
< 35

Days of Rain 0.2mm or more
Autumn

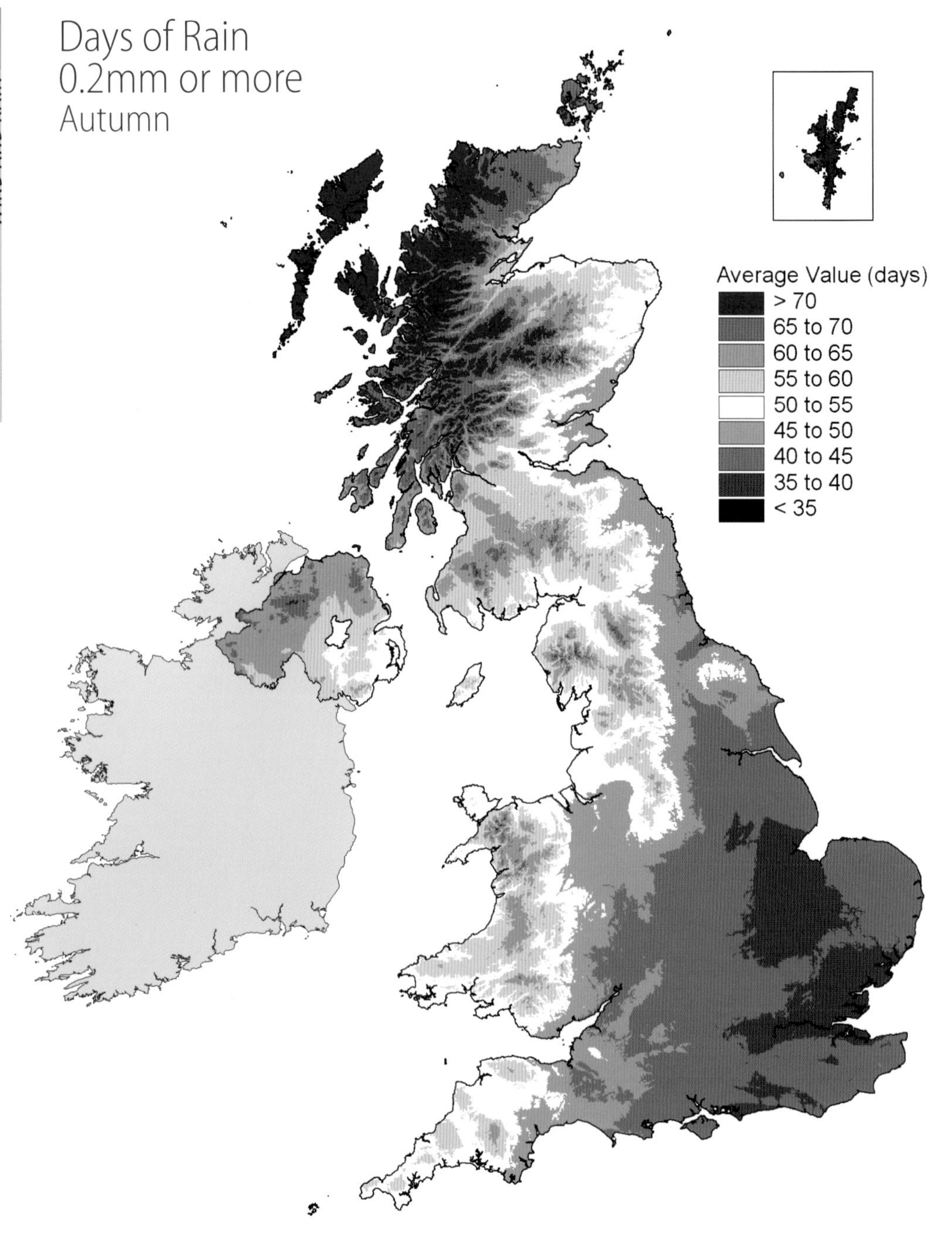
Average Value (days)
> 70
65 to 70
60 to 65
55 to 60
50 to 55
45 to 50
40 to 45
35 to 40
< 35

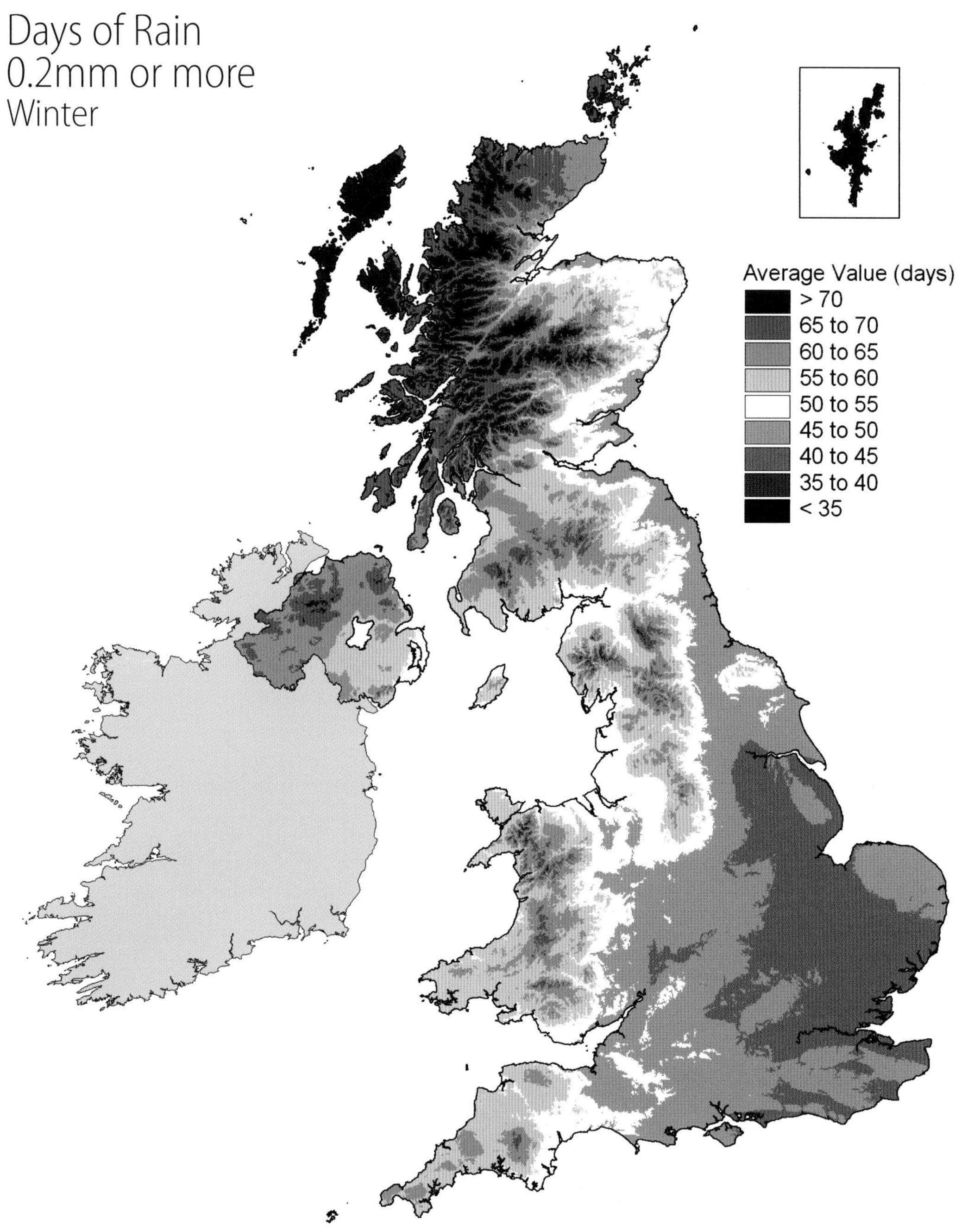
Days of Rain
0.2mm or more
Winter
Average Value (days)
> 70
65 to 70
60 to 65
55 to 60
50 to 55
45 to 50
40 to 45
35 to 40
< 35

Days of Rain 1mm or more
by Season

These maps illustrate the seasonal pattern of 'wet days', which are those when noticeable rain falls to wet the ground. In all seasons there is an east-west split, with less than a third of spring and autumn days bringing wet weather to eastern parts of the country. This drier area expands westwards in summer whereas in winter it becomes confined to parts of eastern England. In winter, it is not unusual for two out of three days in western Scotland to be wet.

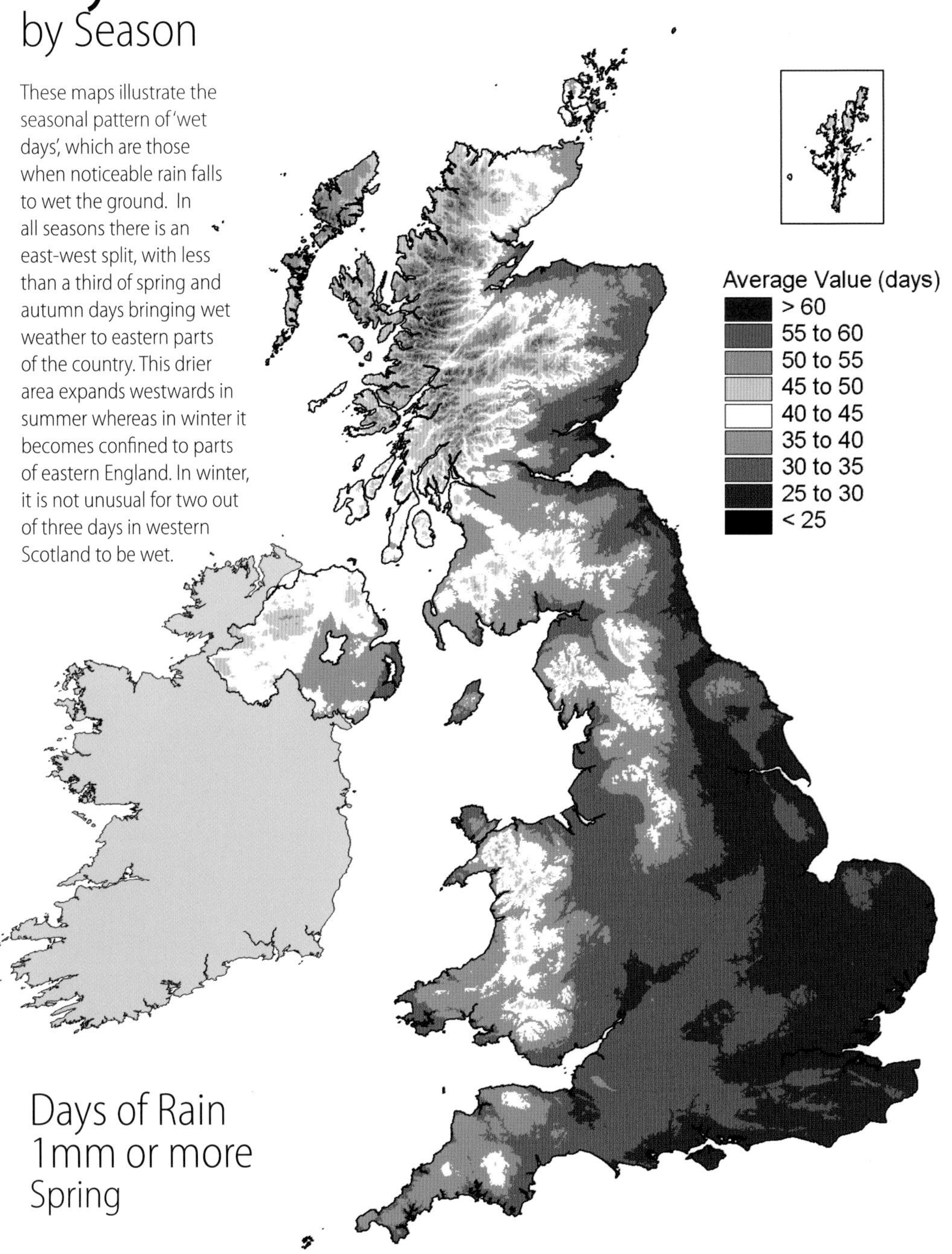
Average Value (days)
> 60
55 to 60
50 to 55
45 to 50
40 to 45
35 to 40
30 to 35
25 to 30
< 25

Days of Rain 1mm or more Spring

Days of Rain 1mm or more
Summer

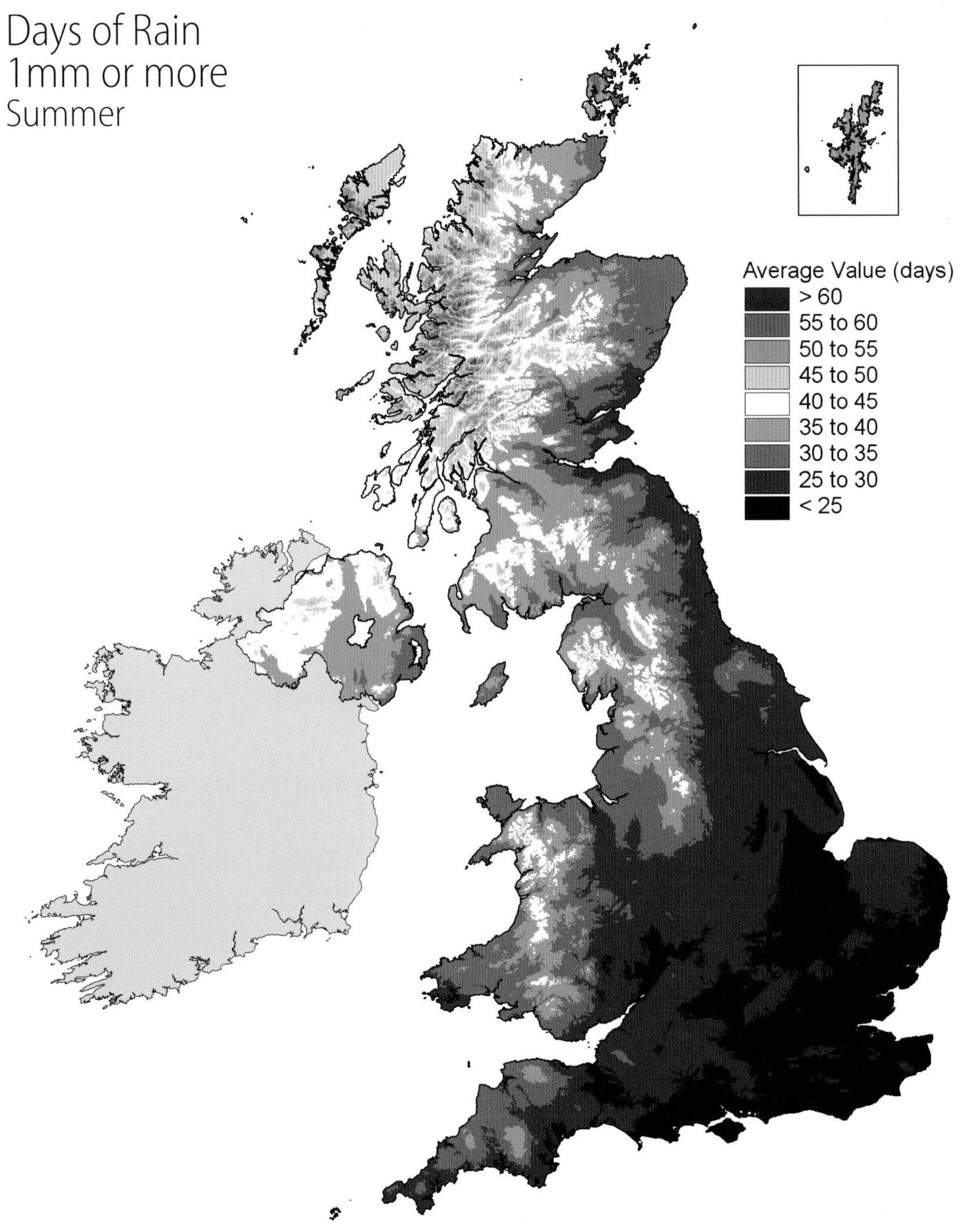
Average Value (days)
> 60
55 to 60
50 to 55
45 to 50
40 to 45
35 to 40
30 to 35
25 to 30
< 25

Days of Rain 1mm or more Autumn

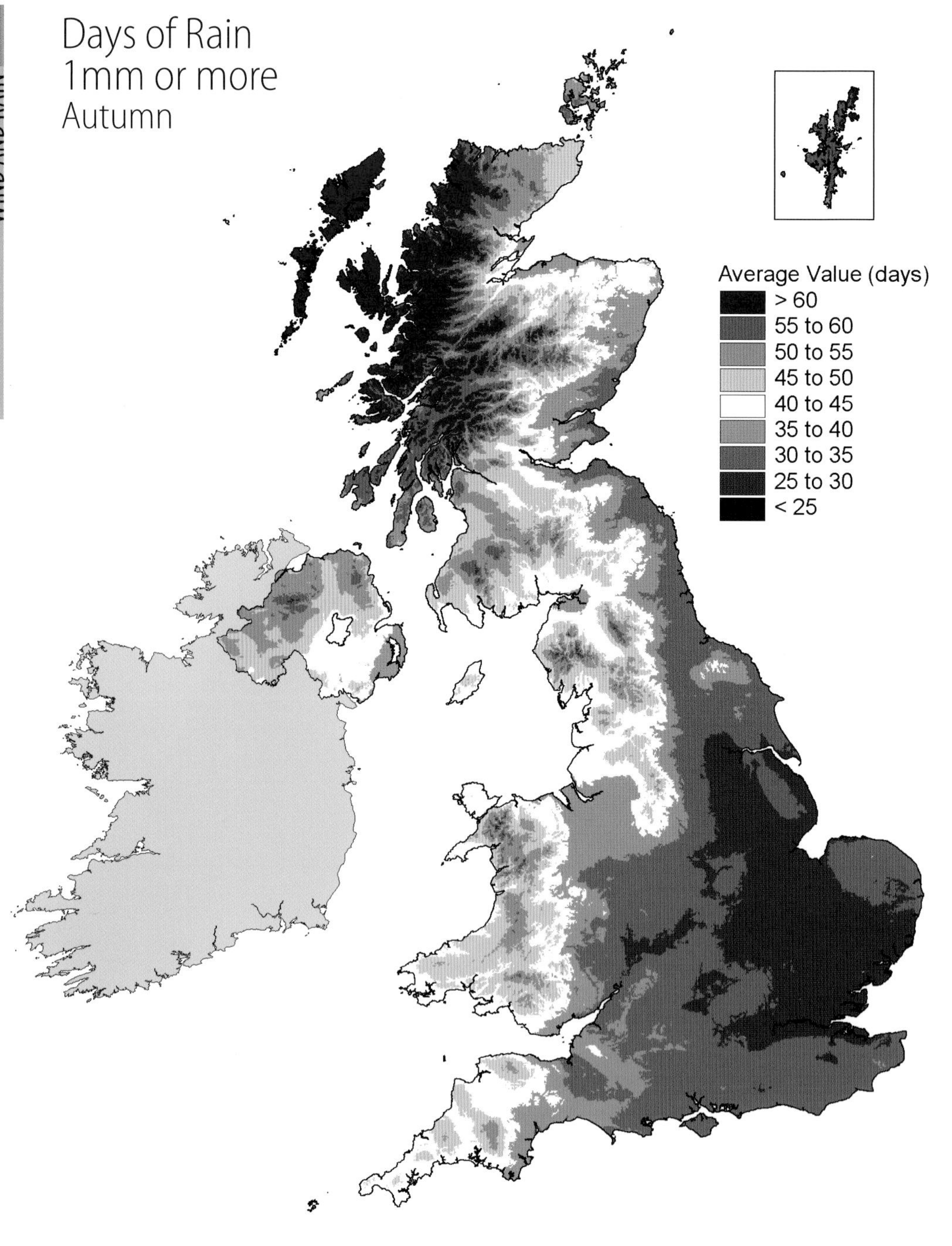
Average Value (days)
> 60
55 to 60
50 to 55
45 to 50
40 to 45
35 to 40
30 to 35
25 to 30
< 25

Days of Rain 1mm or more
Winter

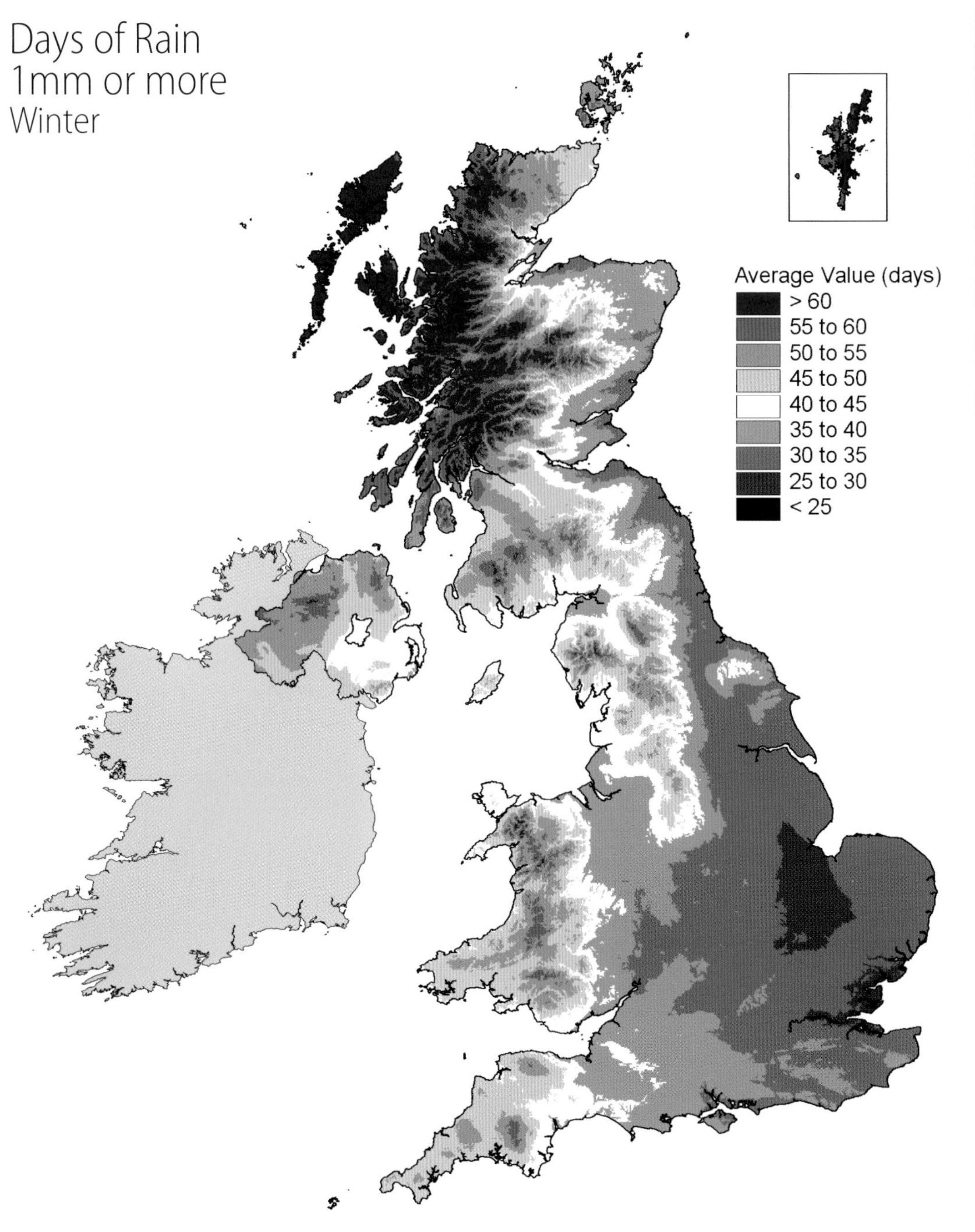

Average Value (days)
> 60
55 to 60
50 to 55
45 to 50
40 to 45
35 to 40
30 to 35
25 to 30
< 25

Days of Rain 10mm or more
by Season

These maps show the seasonal pattern of 'very wet days', which are those with the potential to disrupt outdoor activities and road travel. The variation across the country is striking with only a handful of very wet days in each season in most lowland areas, particularly those well sheltered from the west. Autumn and winter are the seasons when you are most likely to get a soaking, particularly in the western Scottish Highlands where about one day in four is usually very wet.

Average Value (days)

- > 26
- 22 to 26
- 18 to 22
- 14 to 18
- 10 to 14
- 8 to 10
- 6 to 8
- 4 to 6
- < 4

Days of Rain 10mm or more Spring

Days of Rain 10mm or more Summer

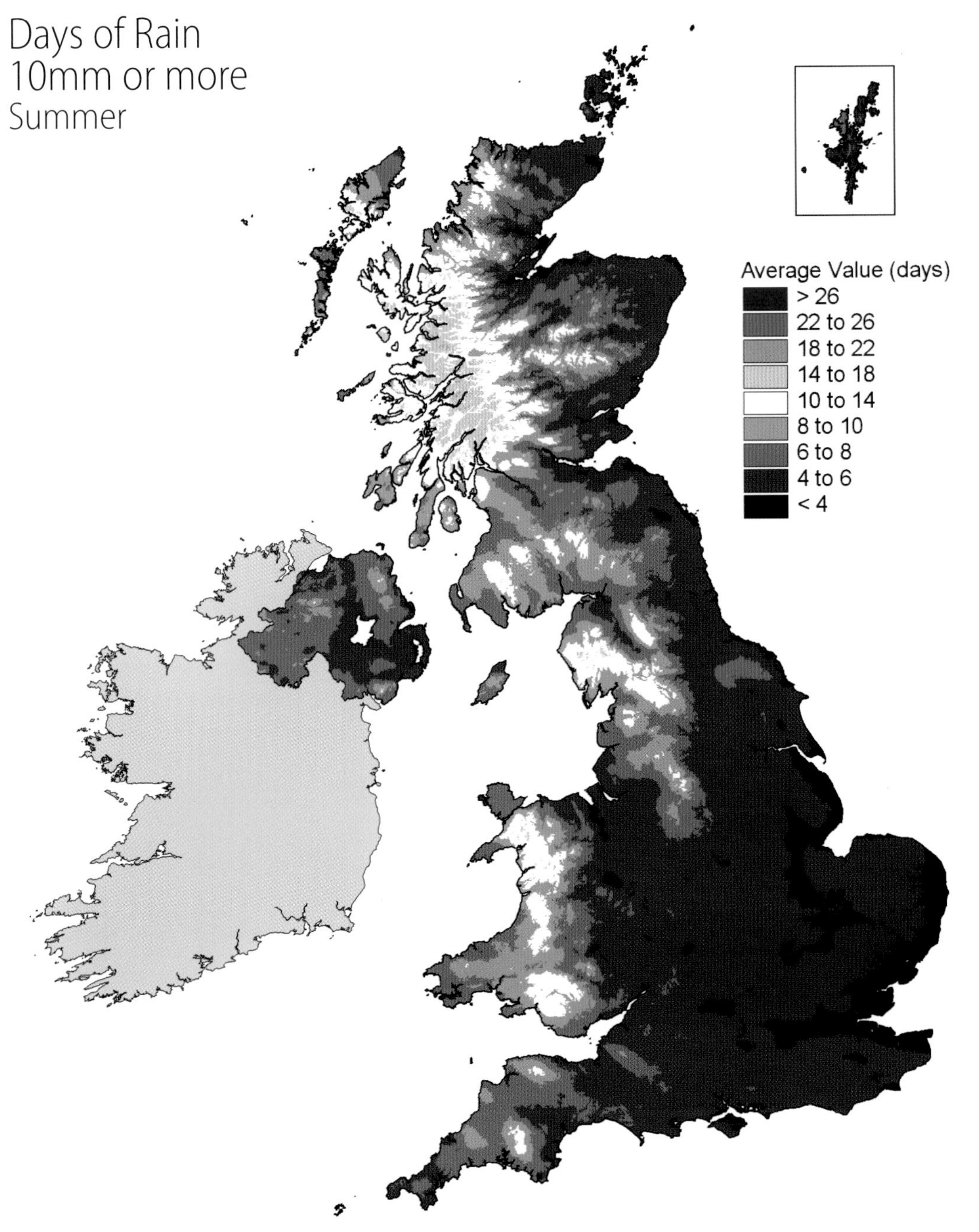

Average Value (days)
> 26
22 to 26
18 to 22
14 to 18
10 to 14
8 to 10
6 to 8
4 to 6
< 4

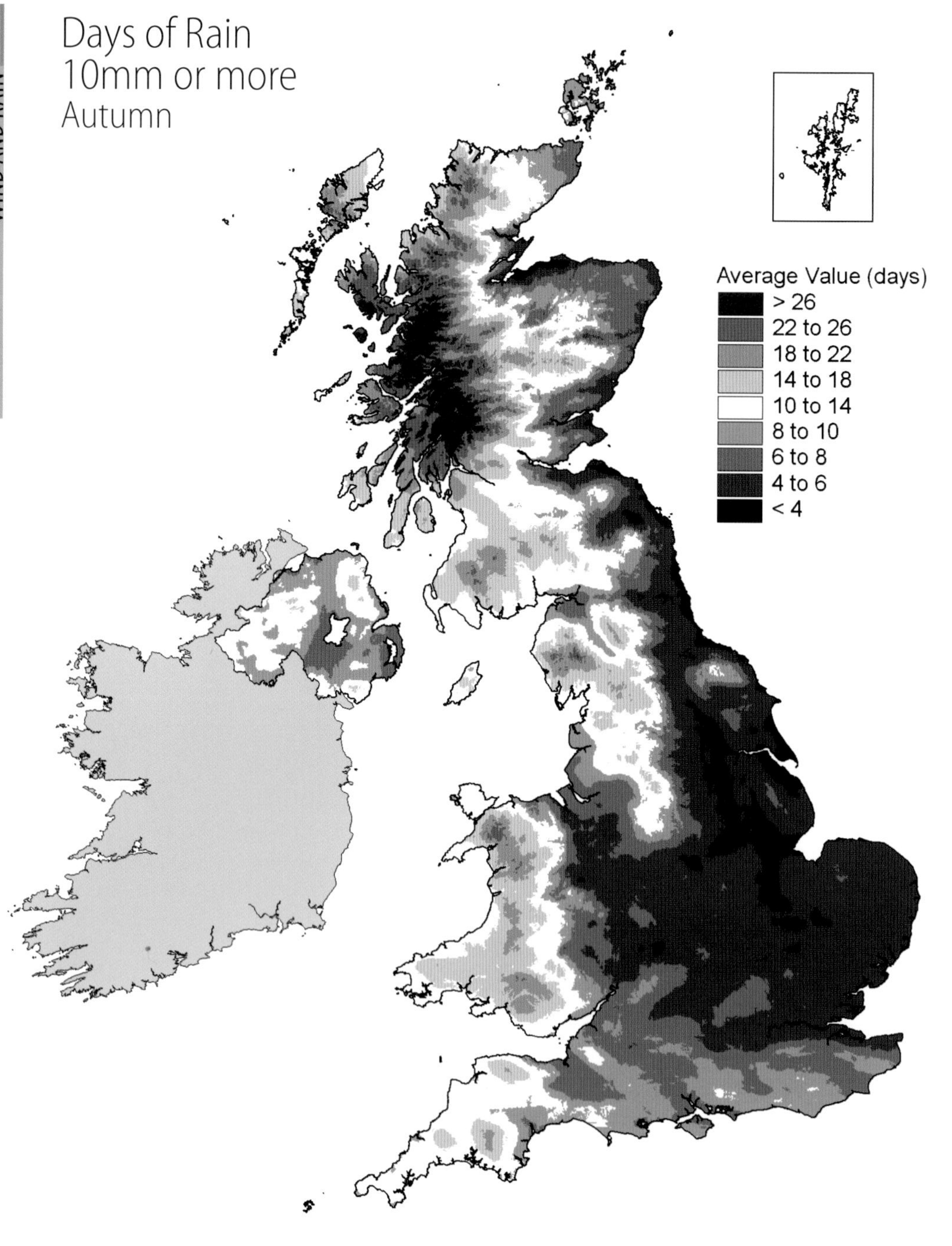
Days of Rain
10mm or more
Autumn
Average Value (days)
> 26
22 to 26
18 to 22
14 to 18
10 to 14
8 to 10
6 to 8
4 to 6
< 4

Days of Rain 10mm or more
Winter

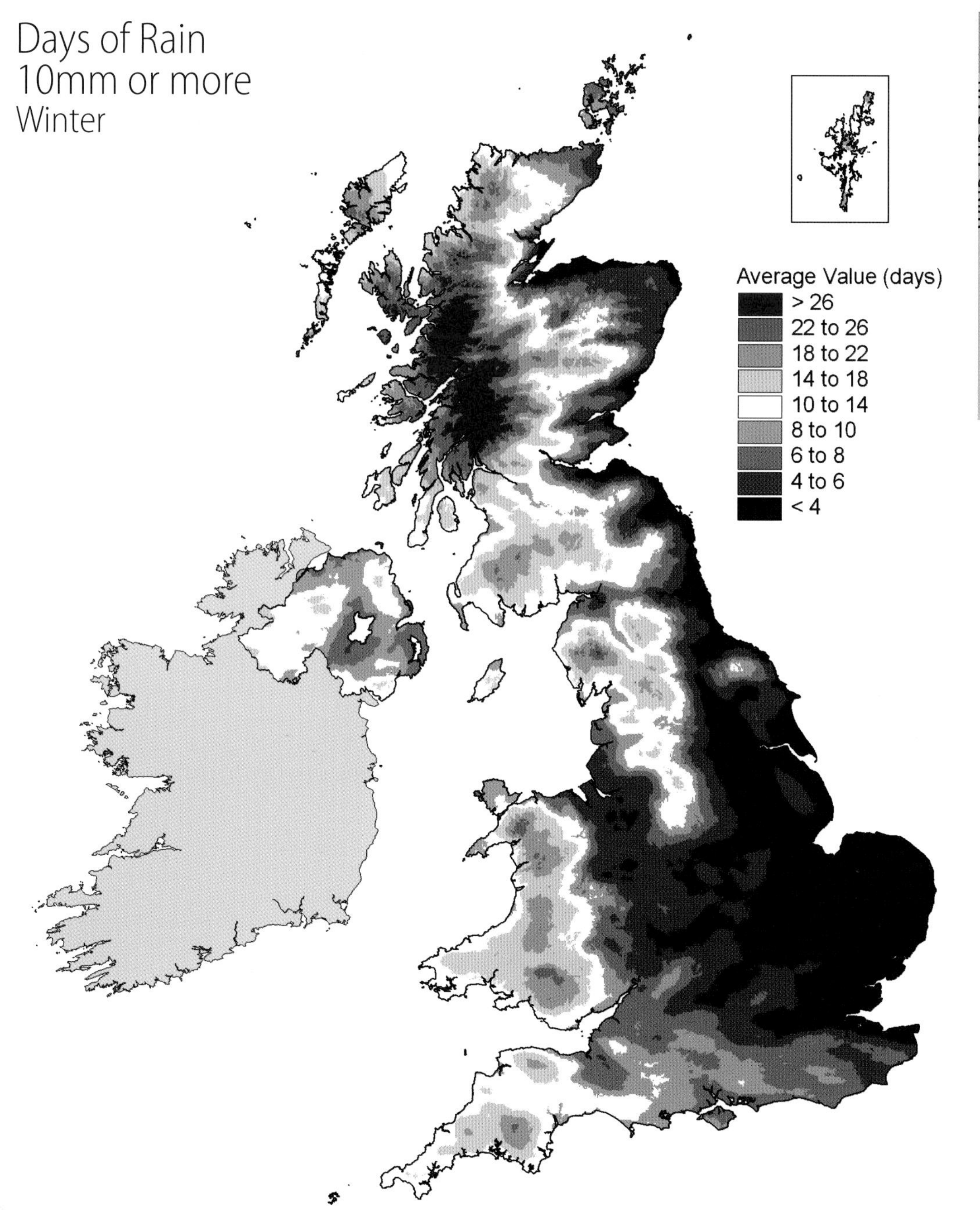

Average Value (days)
> 26
22 to 26
18 to 22
14 to 18
10 to 14
8 to 10
6 to 8
4 to 6
< 4

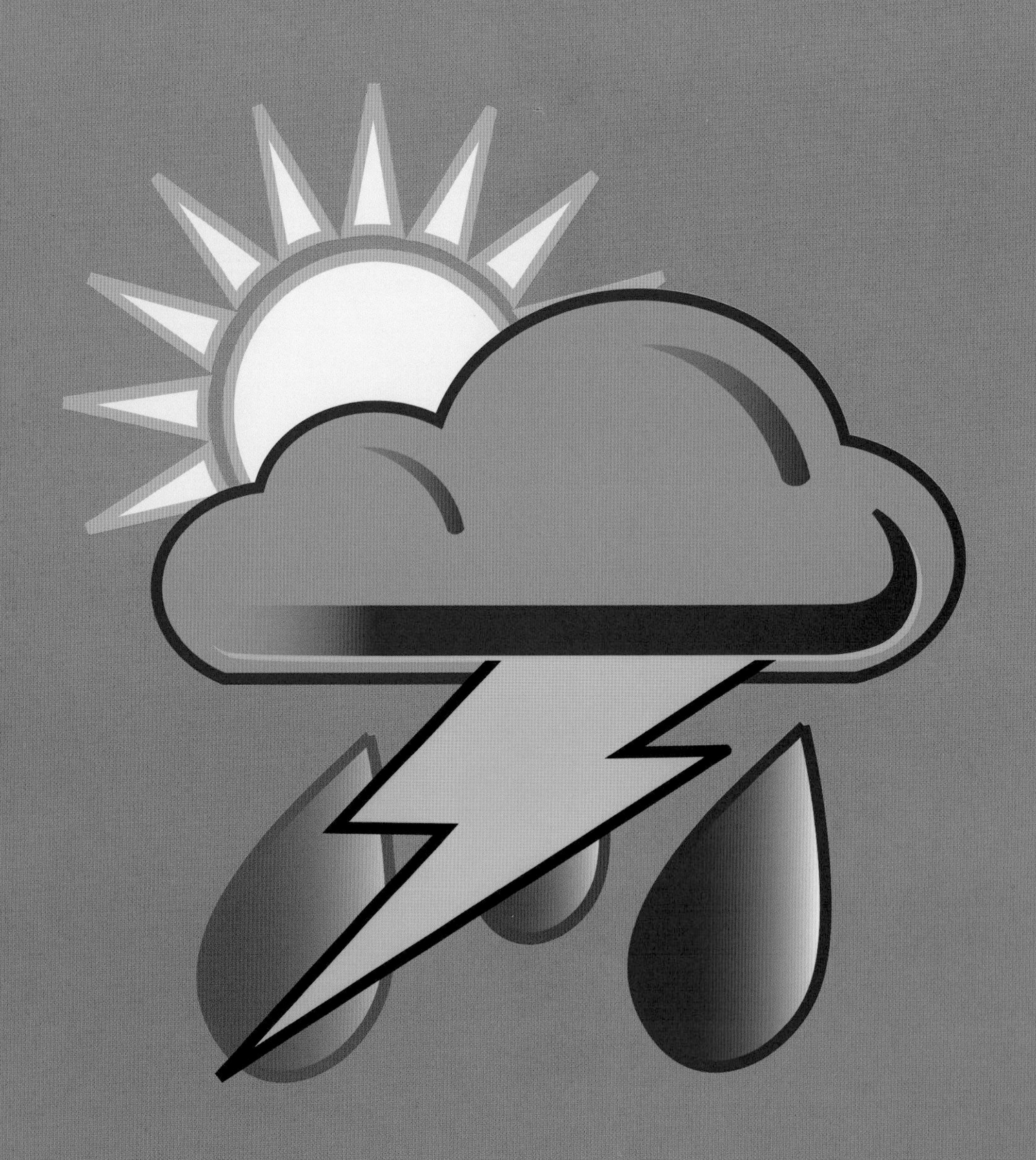

Extreme Weather by Season

'The north wind doth blow, and we shall have snow...'
16th century nursery rhyme

For much of the year Britain is blessed with relatively benign weather. Reports from news stations and national weather services around the world confirm that the most extreme weather occurs elsewhere. However, British weather still has the ability to disrupt our lives – especially affecting travel and outdoor events. Motorists can get stranded in snow and tennis matches at the Wimbledon Championships in late June can be washed out by a thunderstorm. Perhaps it's the relative rarity of extreme weather over most of Britain that means we are sometimes not very well prepared for it. Fortunately, though, any disruption is usually over as quickly as it began.

In contrast, higher upland areas of the country along with most of Scotland have comparatively severe climates and are more used to prolonged spells of frost and snow during winter. In summer, eastern areas of England can see torrential rainfall that wouldn't be out of place in the tropics.

The maps on the following pages show the typical frequency of some of the more extreme weather, from frost and snow to thunder and hail. There can be considerable year-to-year variations, though. For example, early 2009 saw some widespread snowfalls and penetrating frosts after a run of mild winters.

Days of Air Frost

by Season

An air frost occurs when the air temperature falls below freezing point. Upland and rural areas well inland are the most frost-prone, whereas coasts and cities often avoid these freezing conditions. Frost occurs most frequently from late autumn through to spring but summer isn't necessarily frost-free, as Scottish Glens and 'frost hollows' across the country can have the occasional slight frost then.

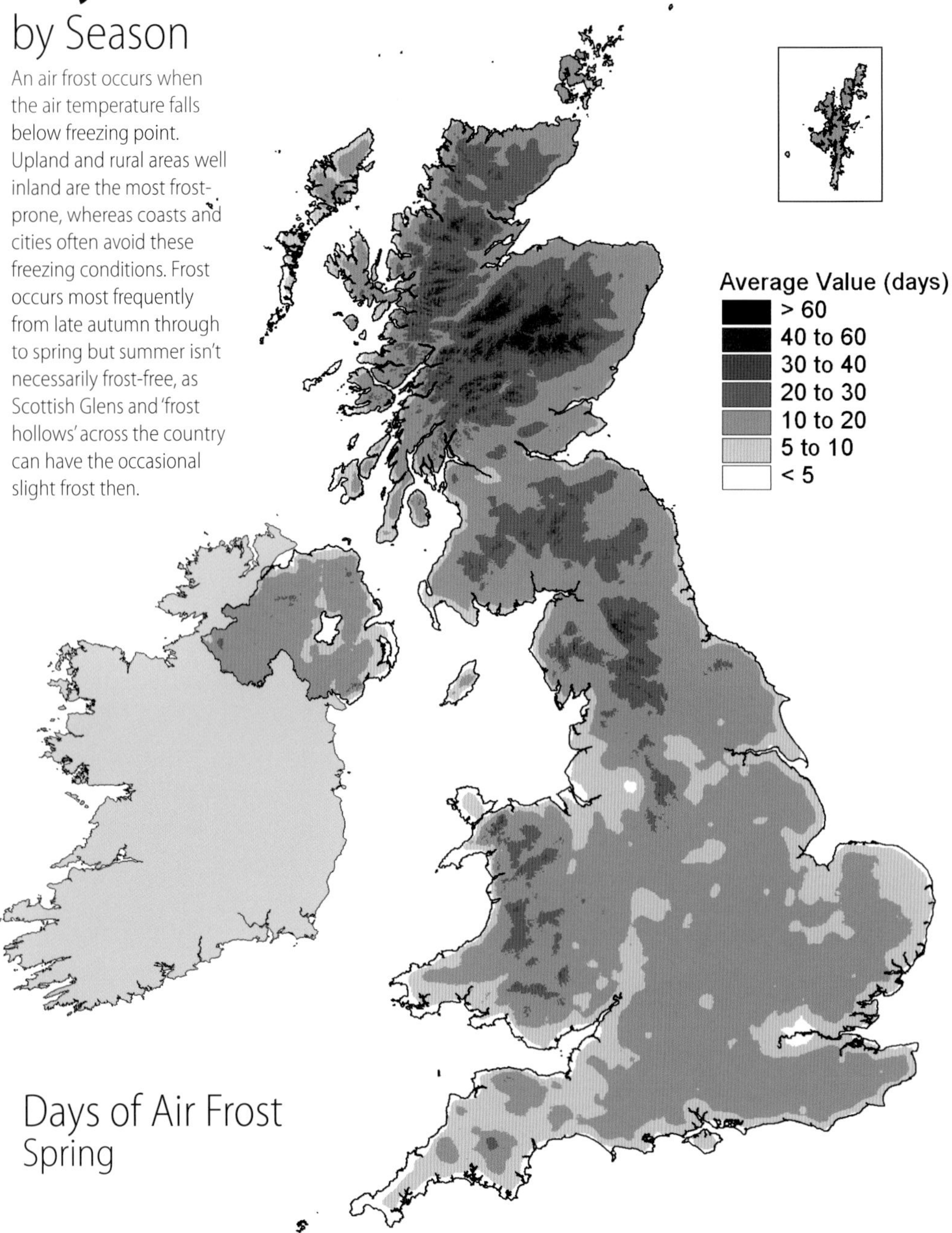
Average Value (days)
> 60
40 to 60
30 to 40
20 to 30
10 to 20
5 to 10
< 5

Days of Air Frost
Spring

Days of Air Frost
Summer

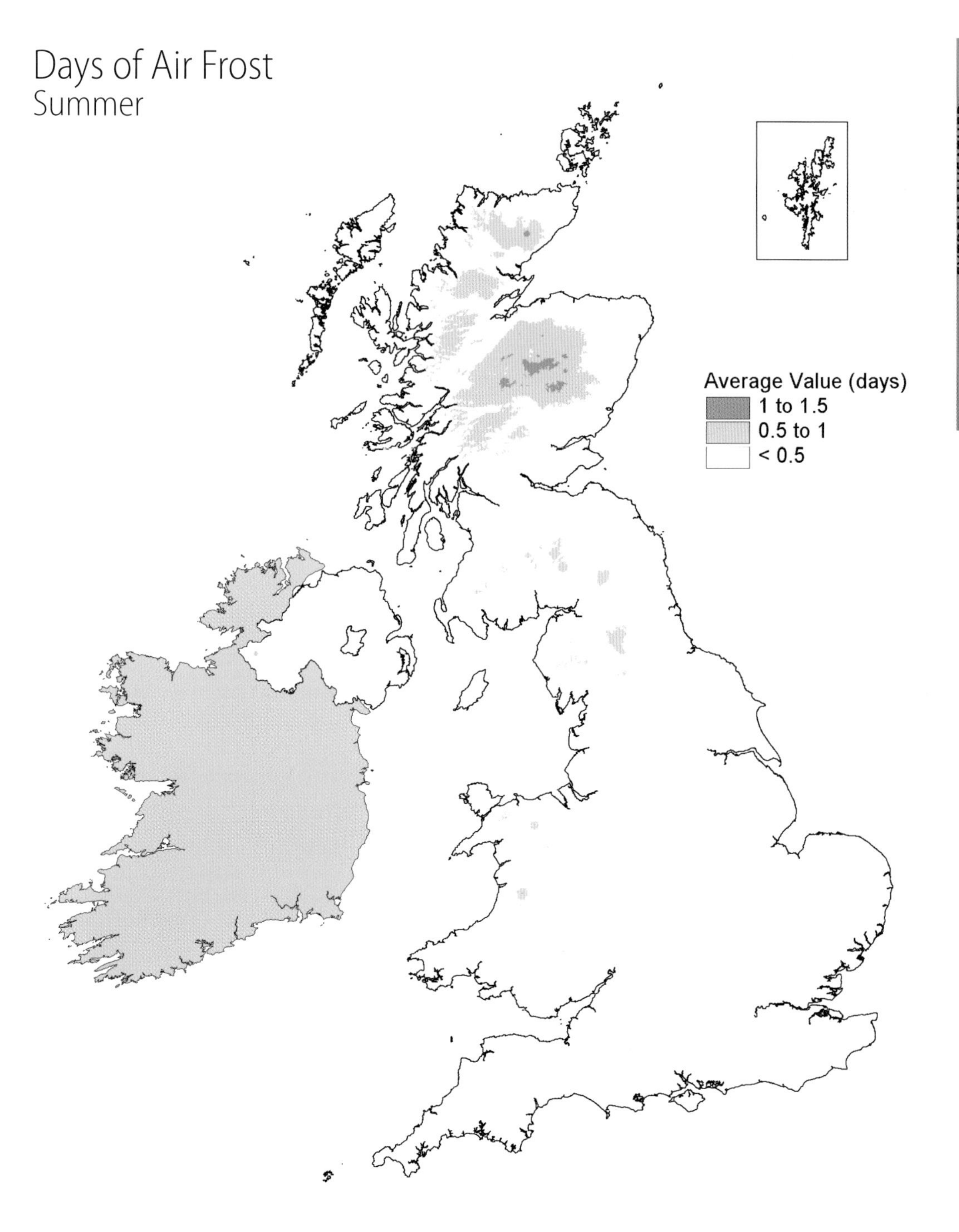
Average Value (days)
1 to 1.5
0.5 to 1
< 0.5

Days of Air Frost
Autumn

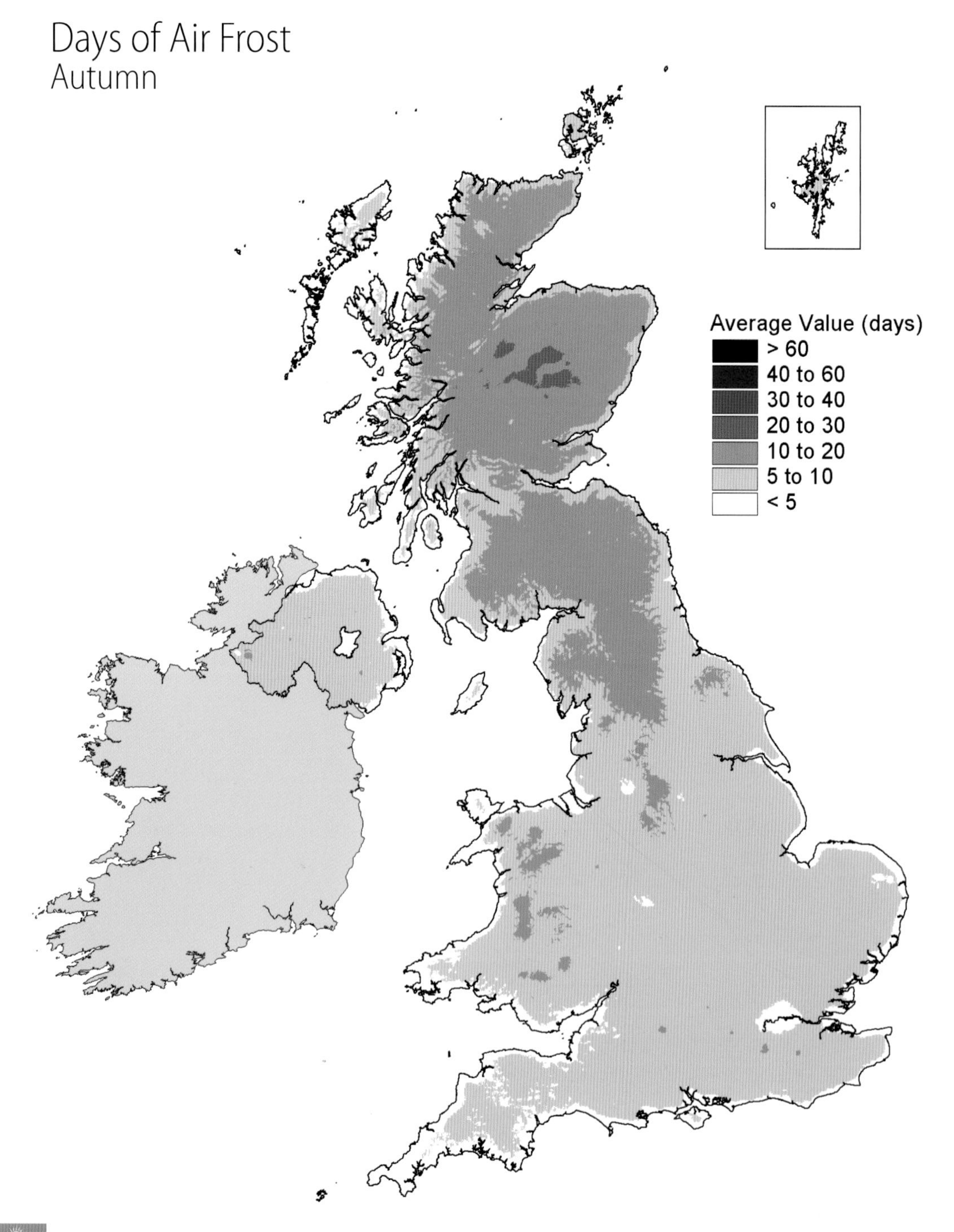
Average Value (days)
> 60
40 to 60
30 to 40
20 to 30
10 to 20
5 to 10
< 5

Days of Air Frost
Winter

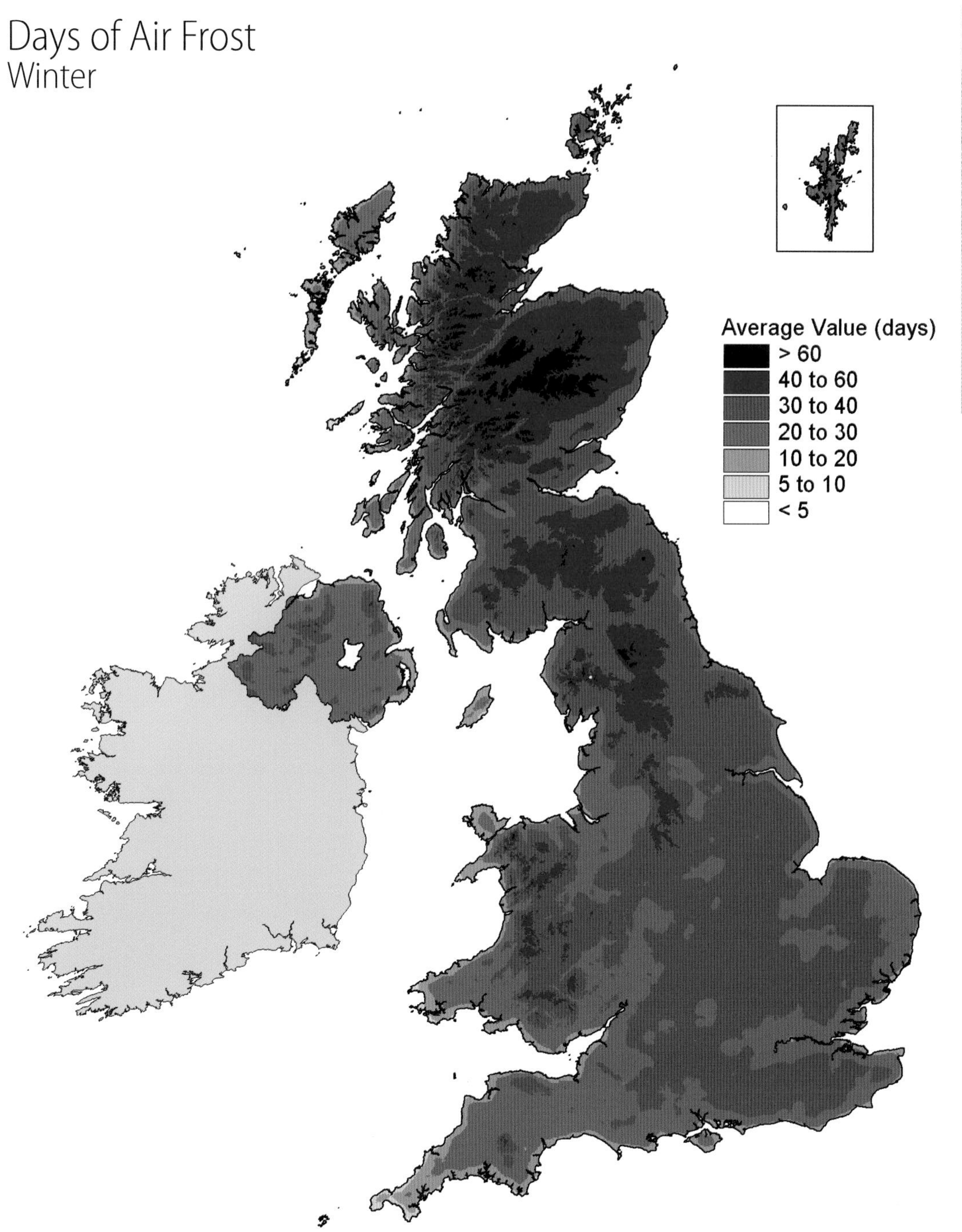
Average Value (days)
> 60
40 to 60
30 to 40
20 to 30
10 to 20
5 to 10
< 5

Days of Ground Frost

by Season

A ground frost occurs when the temperature measured on a grass surface falls below freezing point. This is most likely to occur on clear, calm nights in winter and early spring. The regional and seasonal patterns of ground frost are similar to those of air frost, with coasts and cities least likely to experience one. However, as the ground cools faster than the air there are always more ground frosts than air frosts, and gardeners and growers need to beware – especially in late spring.

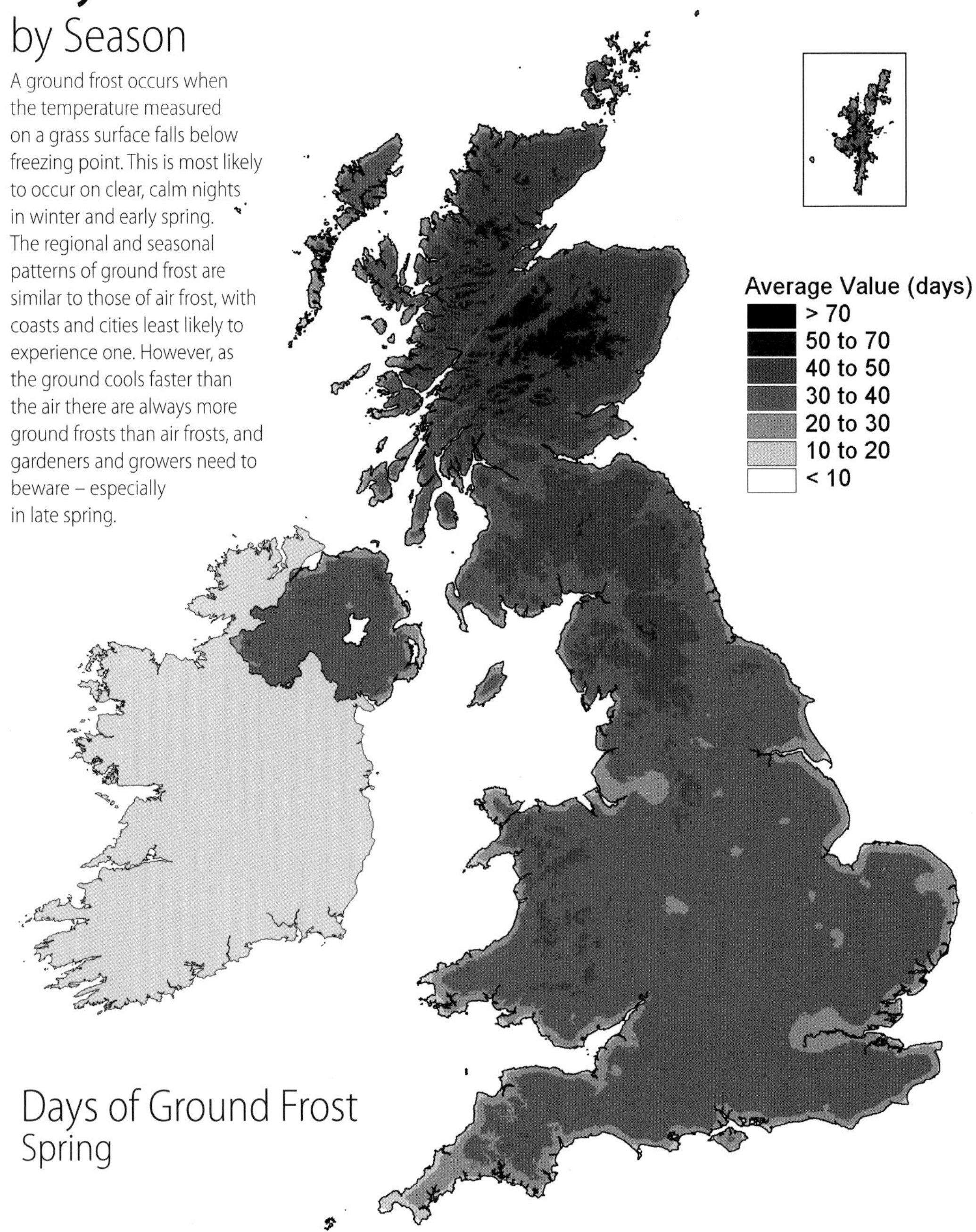
Average Value (days)
> 70
50 to 70
40 to 50
30 to 40
20 to 30
10 to 20
< 10

Days of Ground Frost
Spring

Days of Ground Frost
Summer

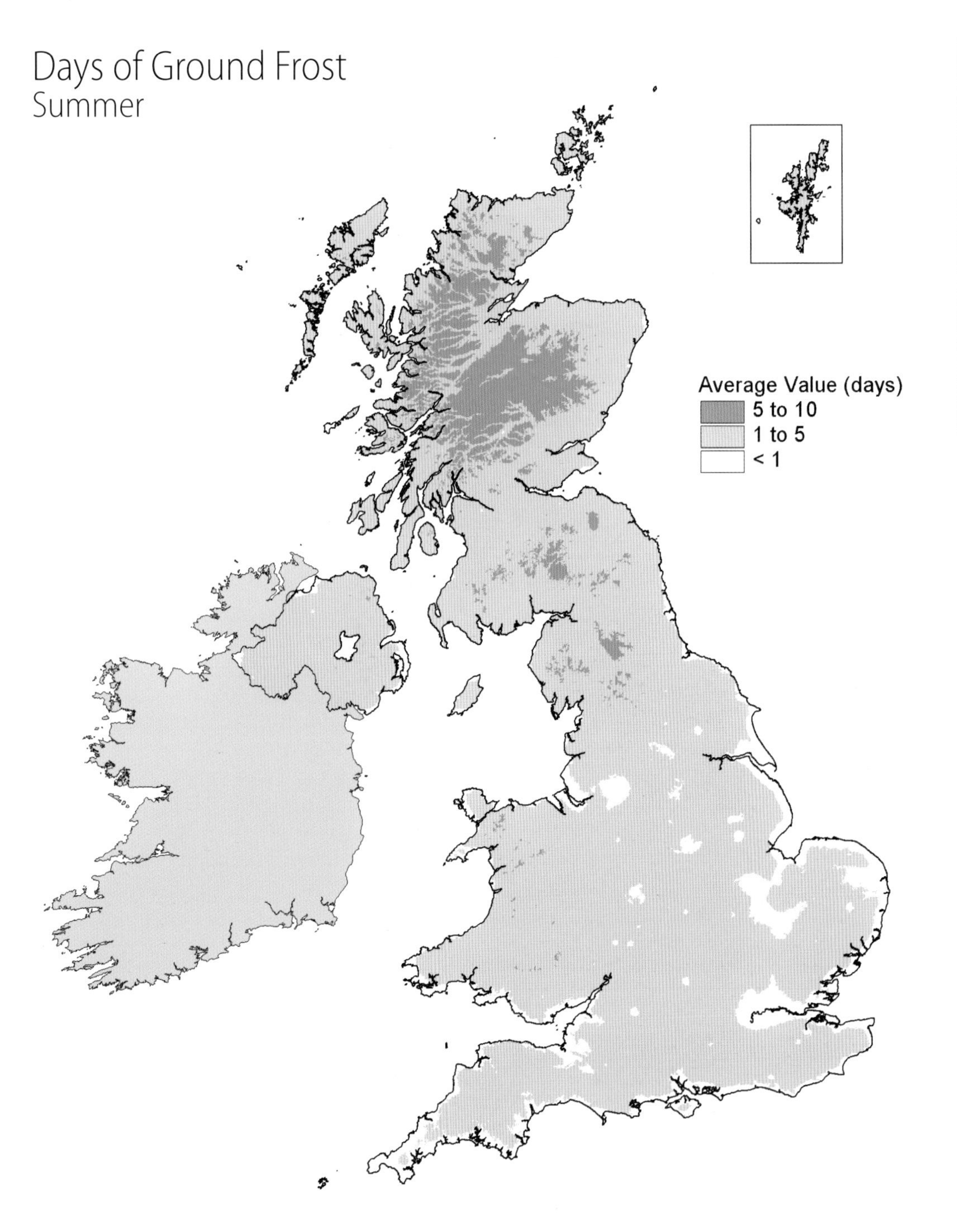
Average Value (days)
5 to 10
1 to 5
< 1

Days of Ground Frost
Autumn

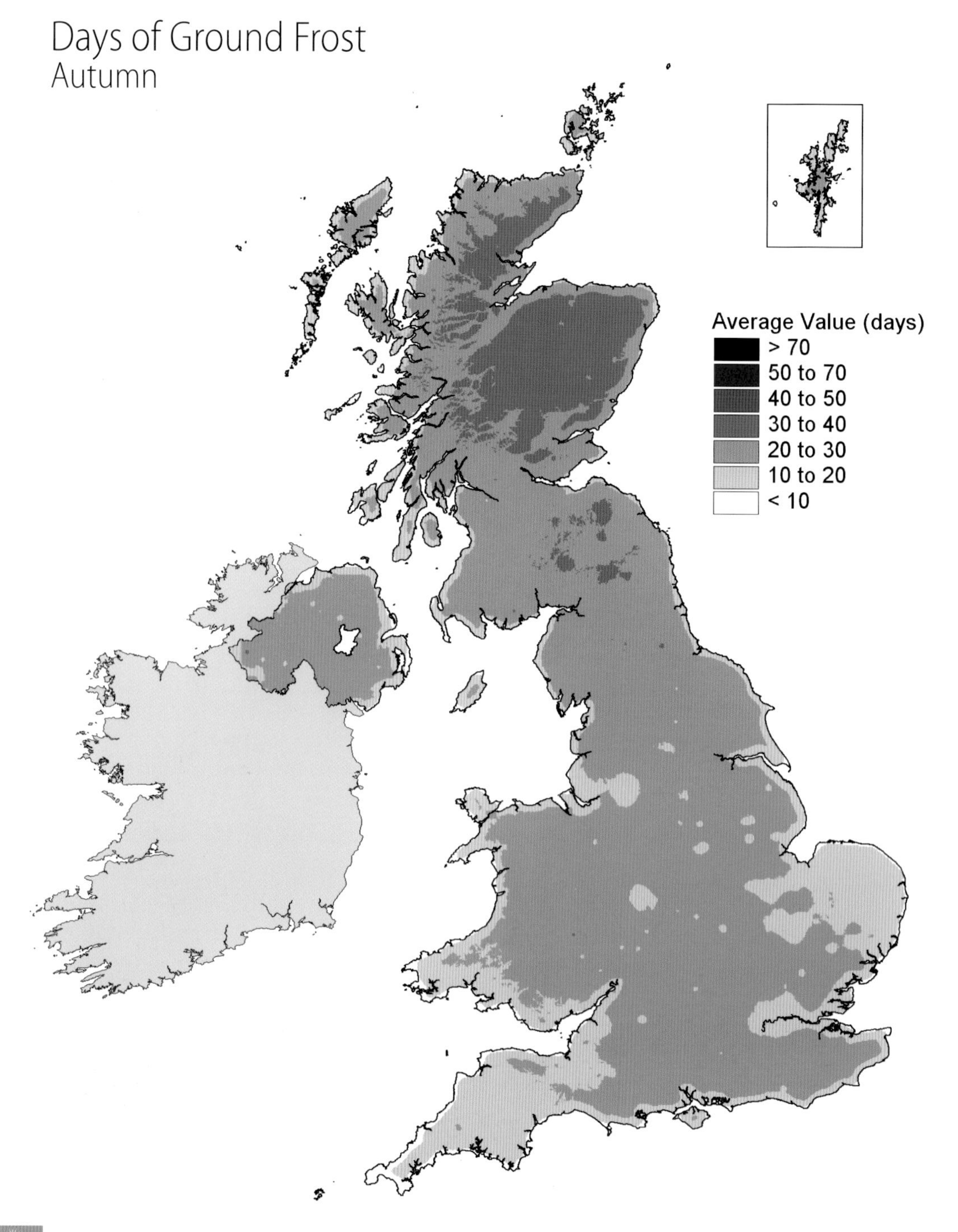
Average Value (days)
> 70
50 to 70
40 to 50
30 to 40
20 to 30
10 to 20
< 10

Days of Ground Frost
Winter

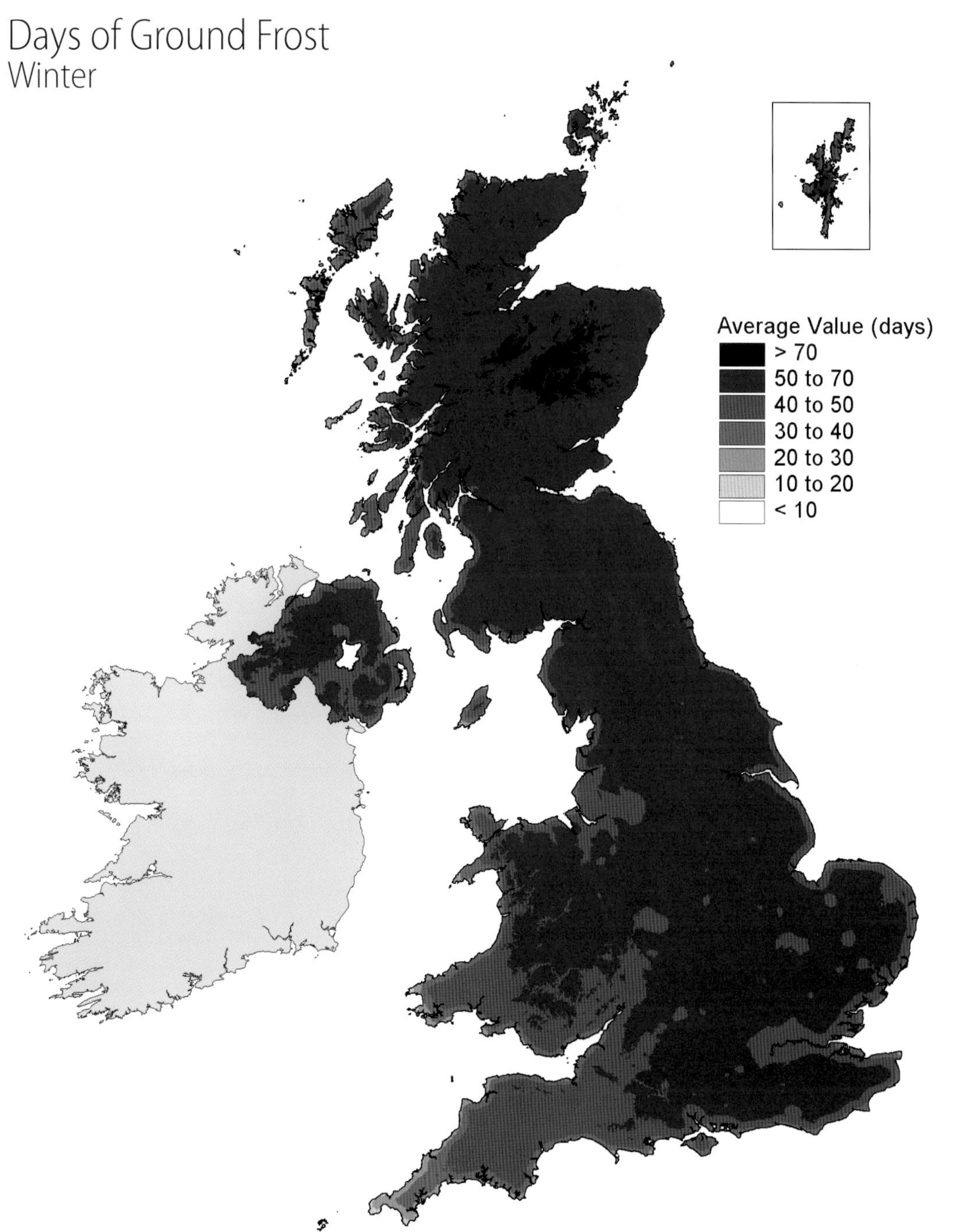

Average Value (days)
> 70
50 to 70
40 to 50
30 to 40
20 to 30
10 to 20
< 10

Days of Sleet/Snow Falling

by Season

The occurrence of sleet and snow is closely linked to temperature, rarely falling when it's higher than 4°C, so the mountains and moors of the north and west are where it's most likely. In the highest areas such as the Cairngorms, it falls on over half the days in winter. Snow is very rare in summer, but not impossible – in early June 1975 snowfall disrupted a county cricket match at Buxton, Derbyshire.

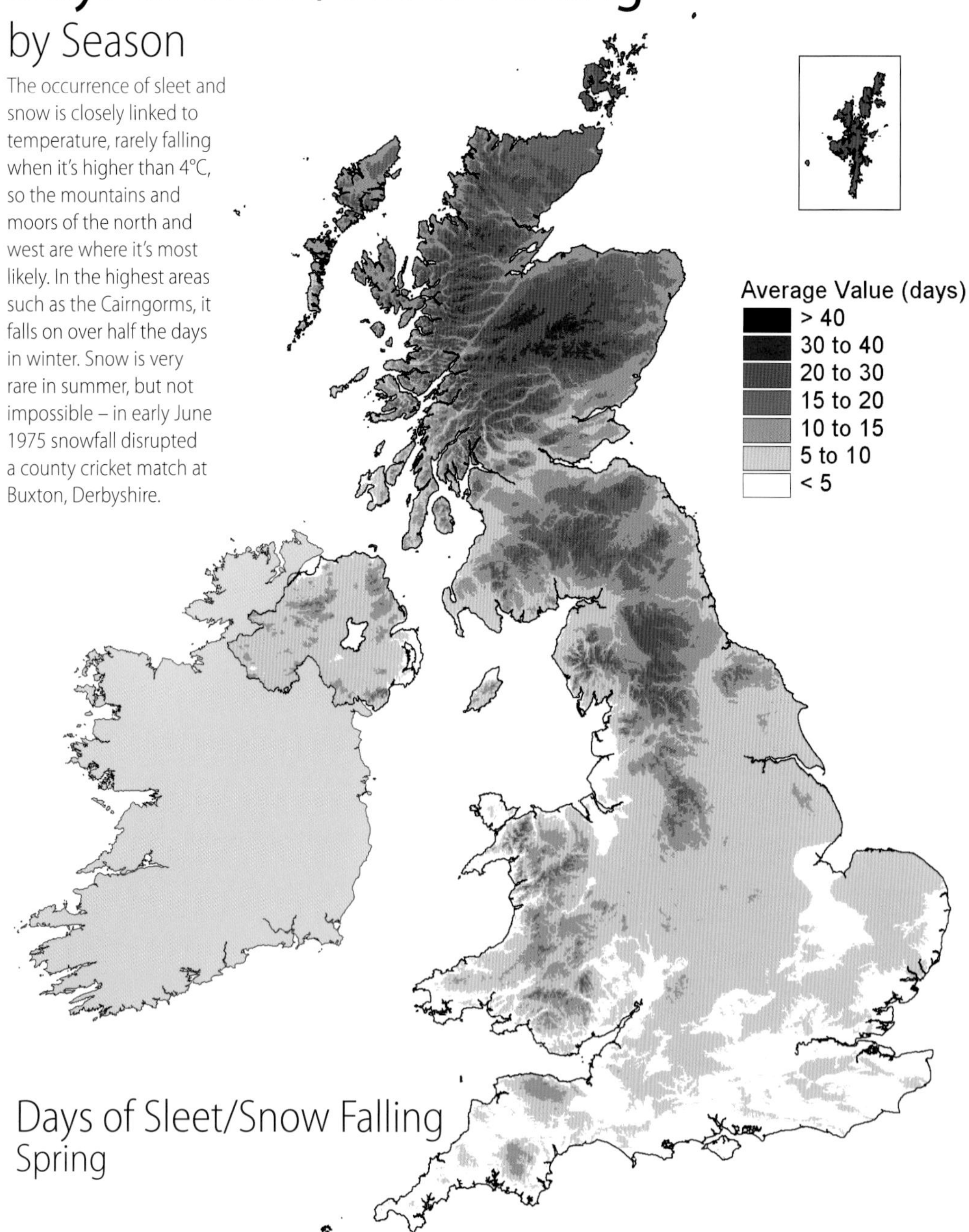
Average Value (days)
> 40
30 to 40
20 to 30
15 to 20
10 to 15
5 to 10
< 5

Days of Sleet/Snow Falling
Spring

Days of Sleet/Snow Falling
Summer

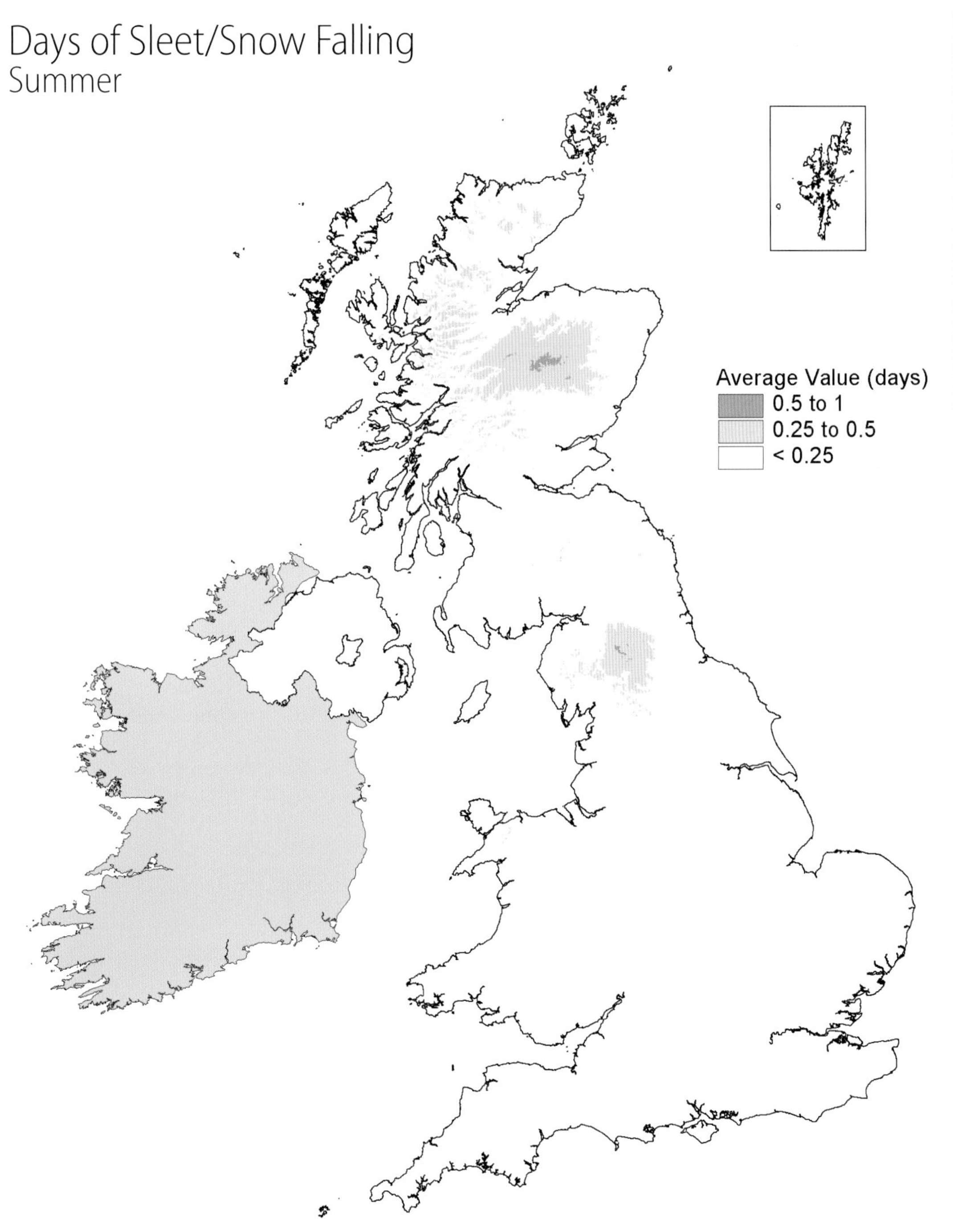
Average Value (days)
0.5 to 1
0.25 to 0.5
< 0.25

Days of Sleet/Snow Falling
Autumn

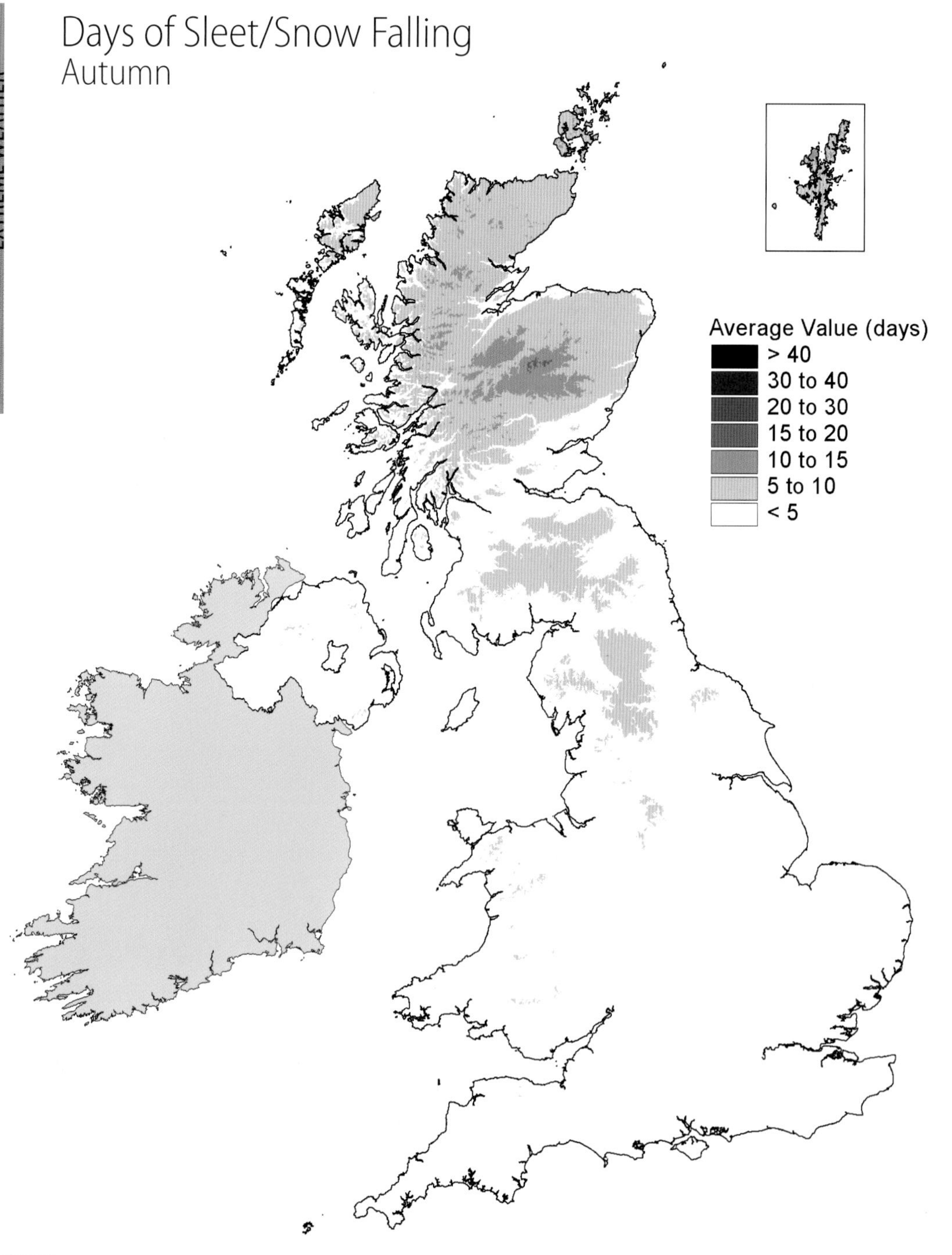
Average Value (days)
> 40
30 to 40
20 to 30
15 to 20
10 to 15
5 to 10
< 5

Days of Sleet/Snow Falling
Winter

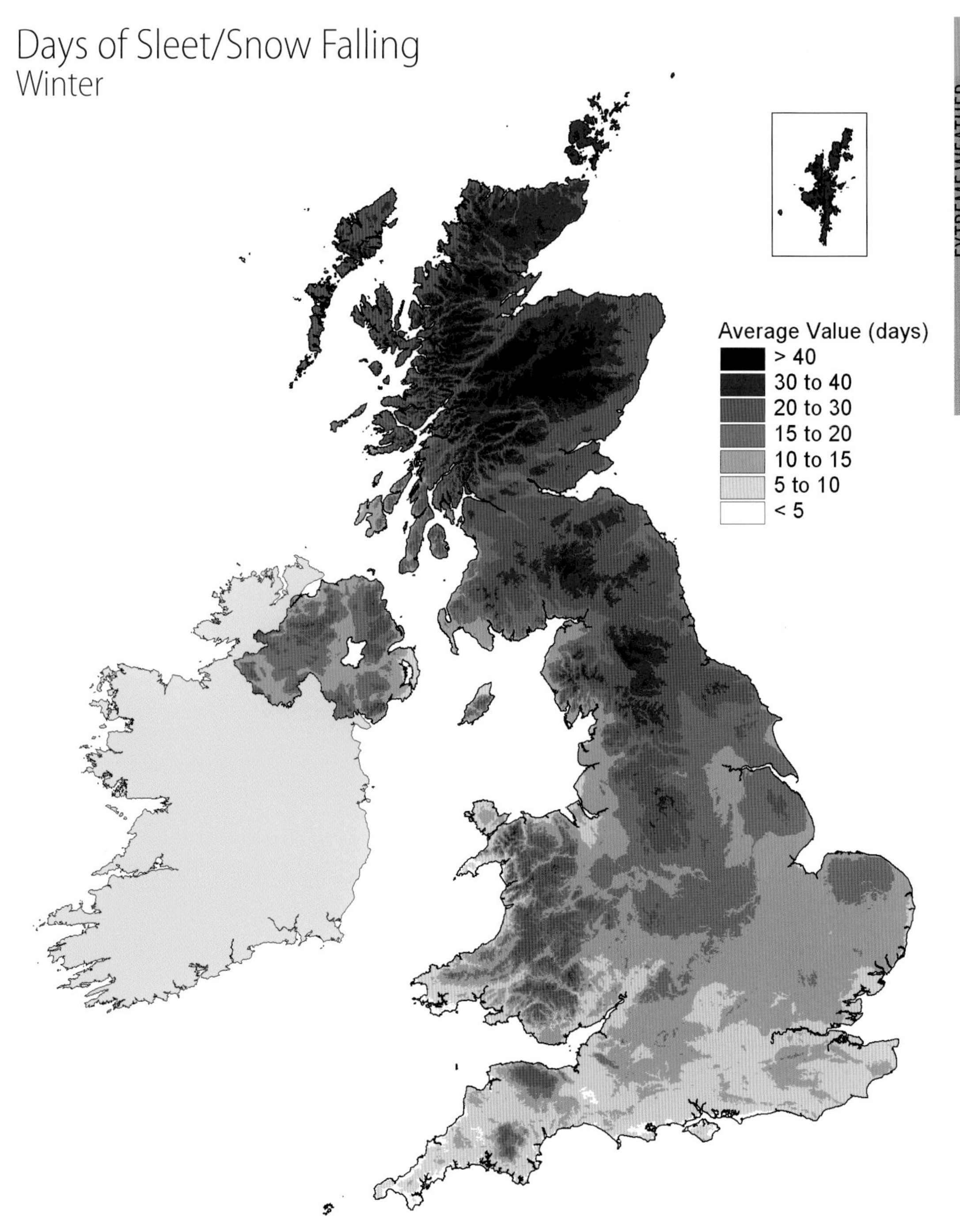
Average Value (days)
> 40
30 to 40
20 to 30
15 to 20
10 to 15
5 to 10
< 5

Days of Snow Lying
by Season

If a place has snow covering over half the ground at 0900 GMT (Greenwich Mean Time), then that day is counted as one with 'lying snow'. The mountains and moors of the north are where snow is most likely to lie, leading to the development of the Scottish ski centres in the Cairngorms and near Ben Nevis. In the severe winter of 1963 it was the turn of England and Wales to have stubborn snow cover, which persisted for over two months in many places.

Average Value (days)
> 40
30 to 40
20 to 30
15 to 20
10 to 15
5 to 10
< 5

Days of Snow Lying
Spring

Days of Snow Lying
Summer

Average Value (days)
< 0.5

Days of Snow Lying
Autumn

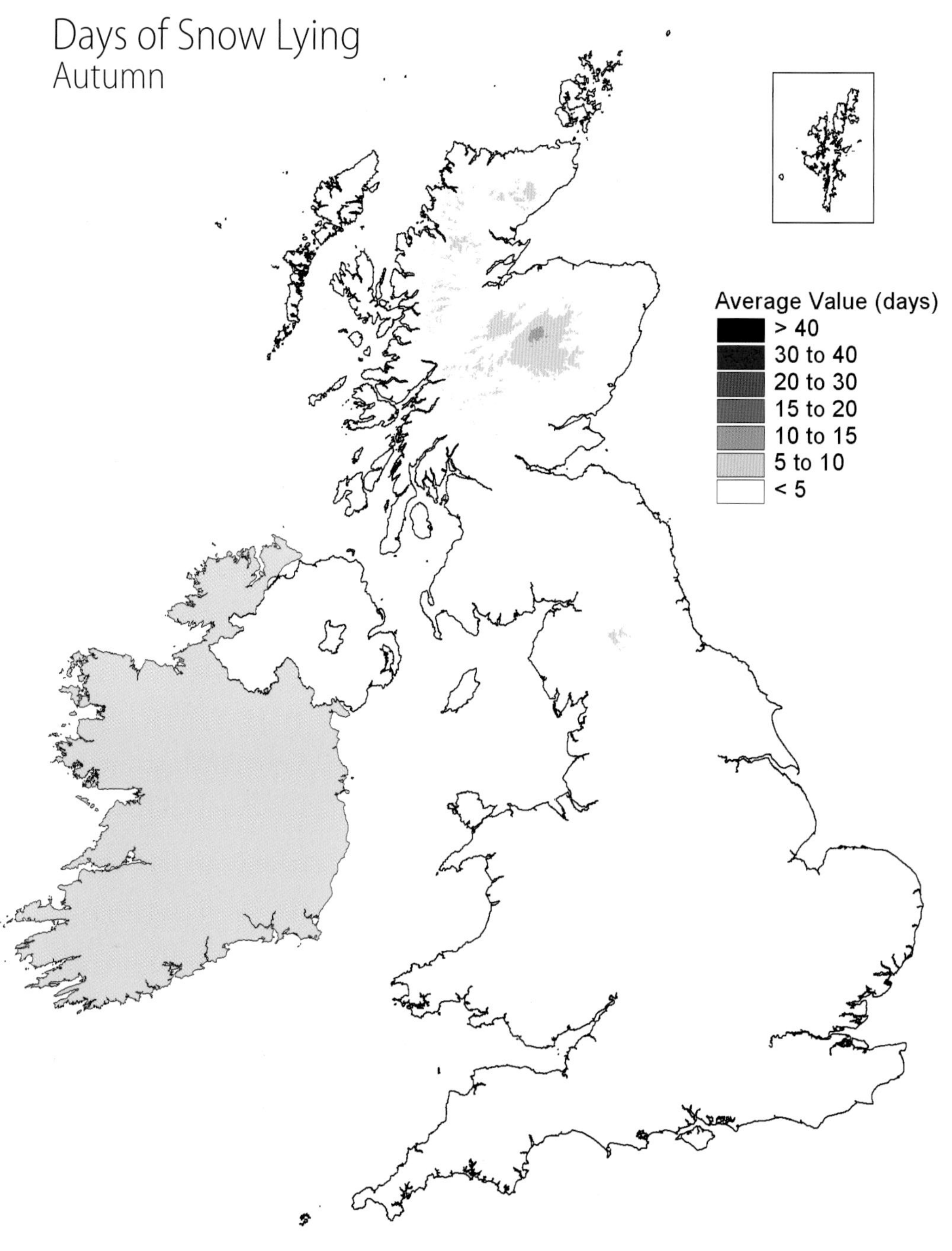
Average Value (days)
> 40
30 to 40
20 to 30
15 to 20
10 to 15
5 to 10
< 5

Days of Snow Lying
Winter

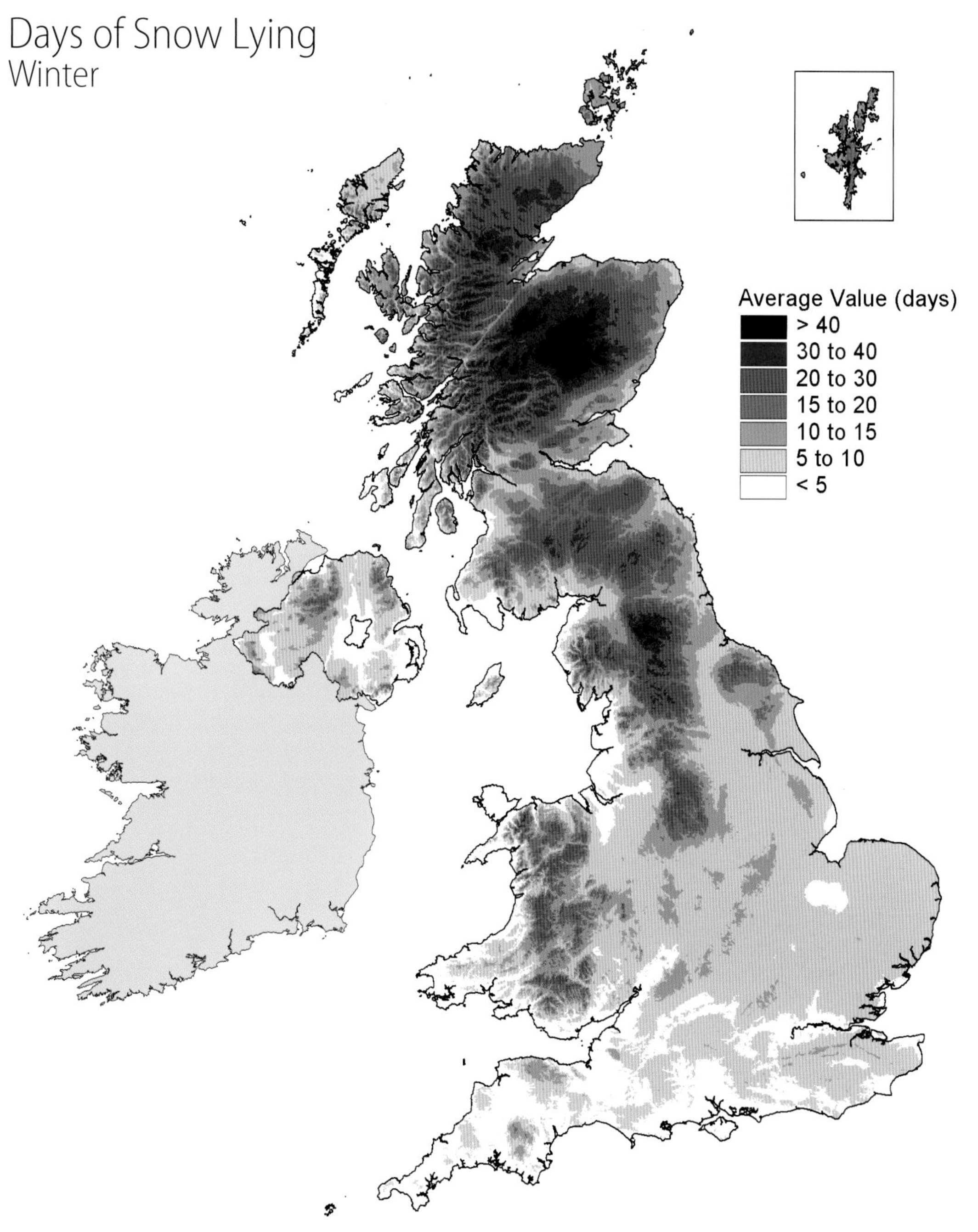

Average Value (days)
> 40
30 to 40
20 to 30
15 to 20
10 to 15
5 to 10
< 5

Days of Thunder
by Season

These maps show the seasonal variation in the number of days when thunder can be heard. They confirm that thunderstorms occur most often in eastern England in the summer and rarely over most of Scotland at any season. Timing the interval between seeing a lightning strike and hearing thunder can help to estimate how far away a storm is, with every five seconds representing a mile. Thunder can be heard up to 20 miles away.

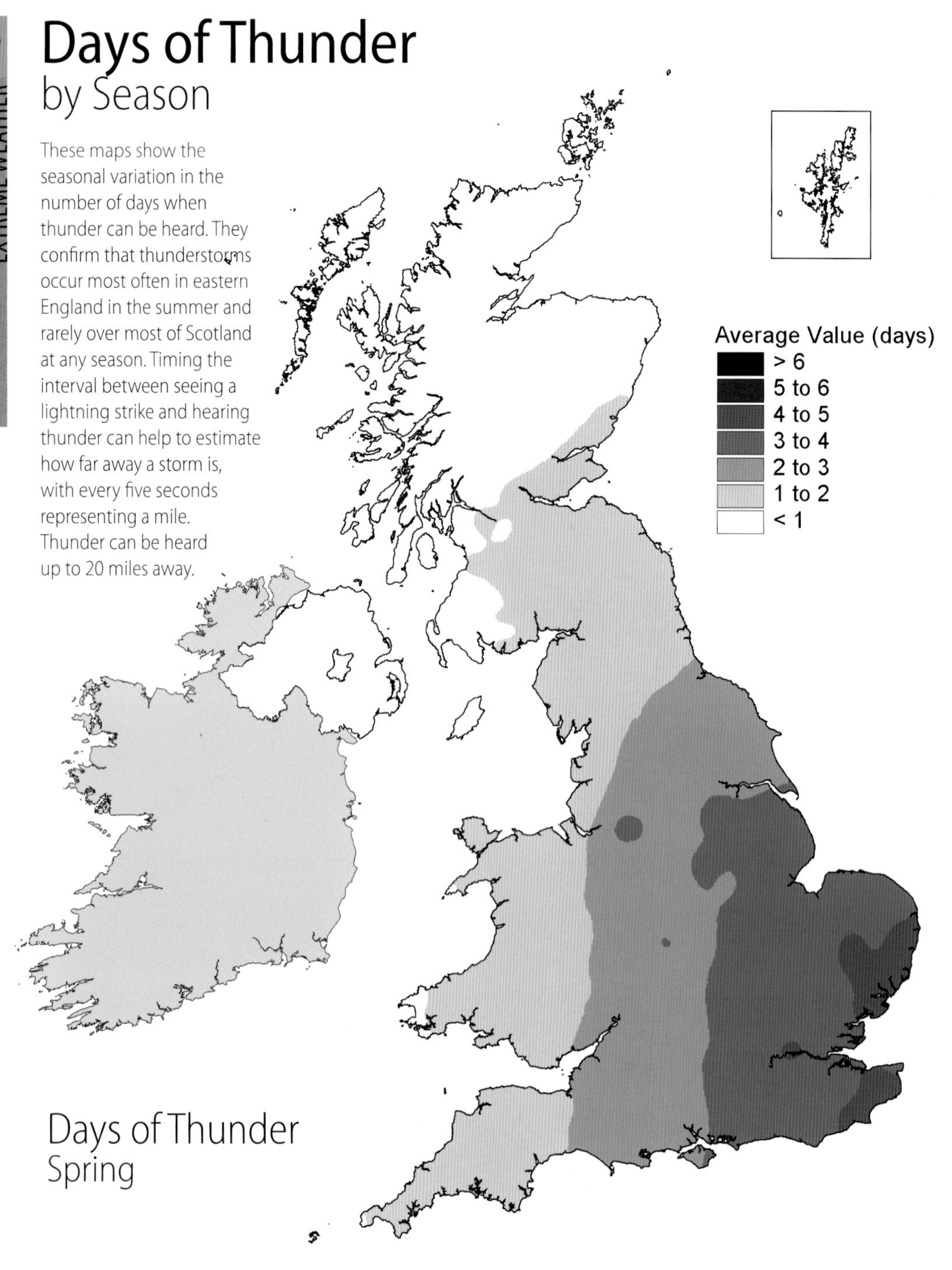
Average Value (days)
> 6
5 to 6
4 to 5
3 to 4
2 to 3
1 to 2
< 1

Days of Thunder
Spring

Days of Thunder
Summer

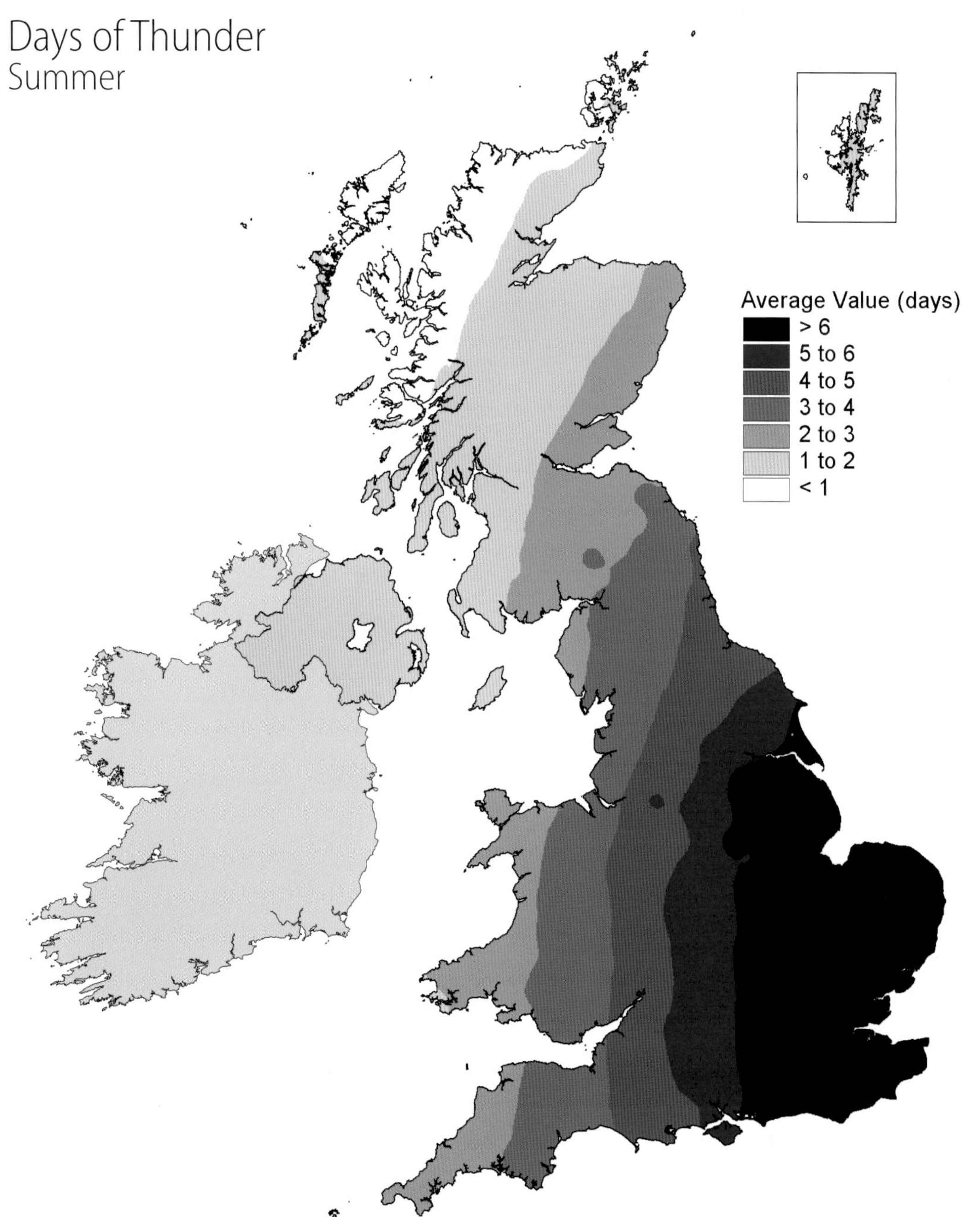

Average Value (days)
> 6
5 to 6
4 to 5
3 to 4
2 to 3
1 to 2
< 1

Days of Thunder
Autumn

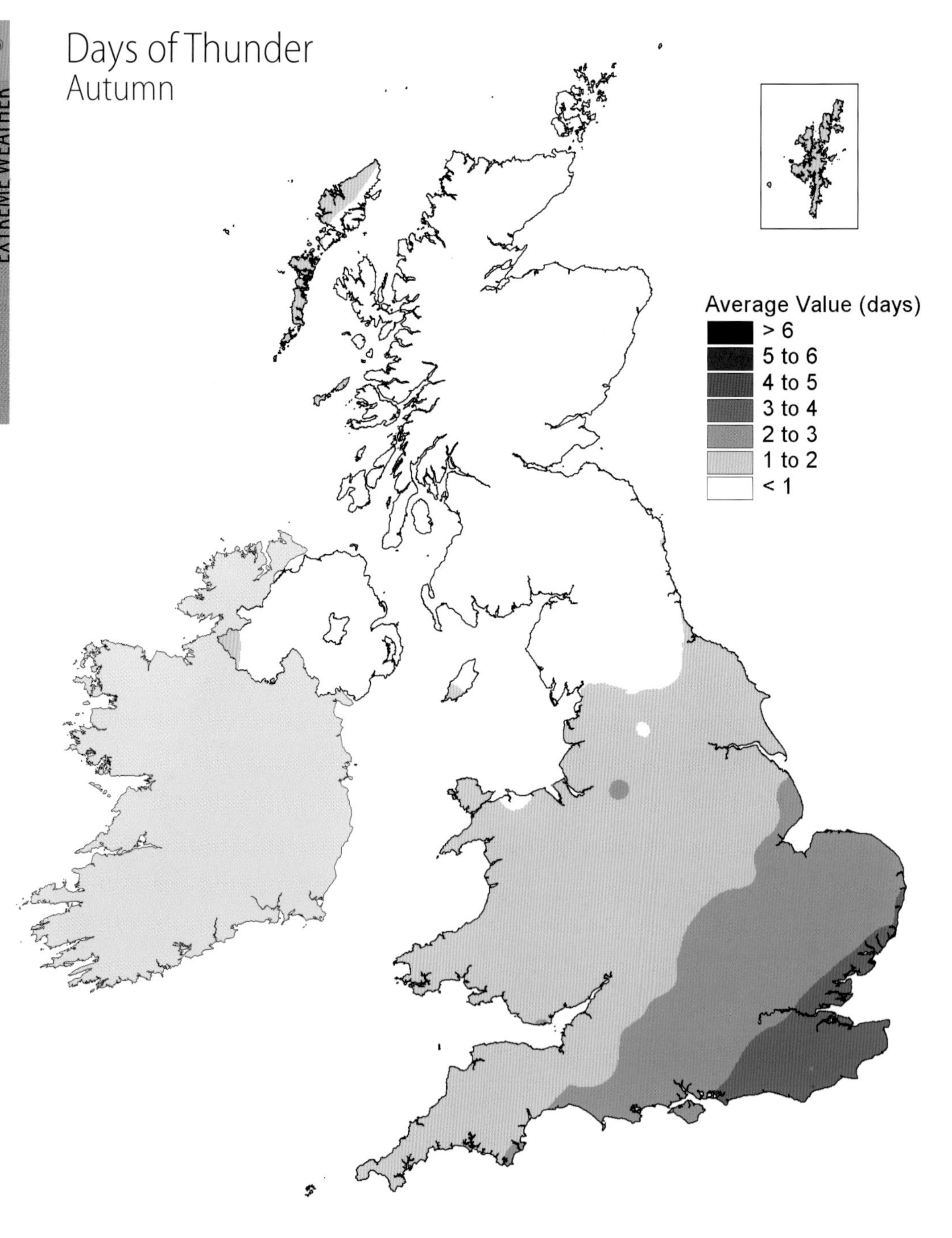
Average Value (days)
> 6
5 to 6
4 to 5
3 to 4
2 to 3
1 to 2
< 1

Days of Thunder
Winter

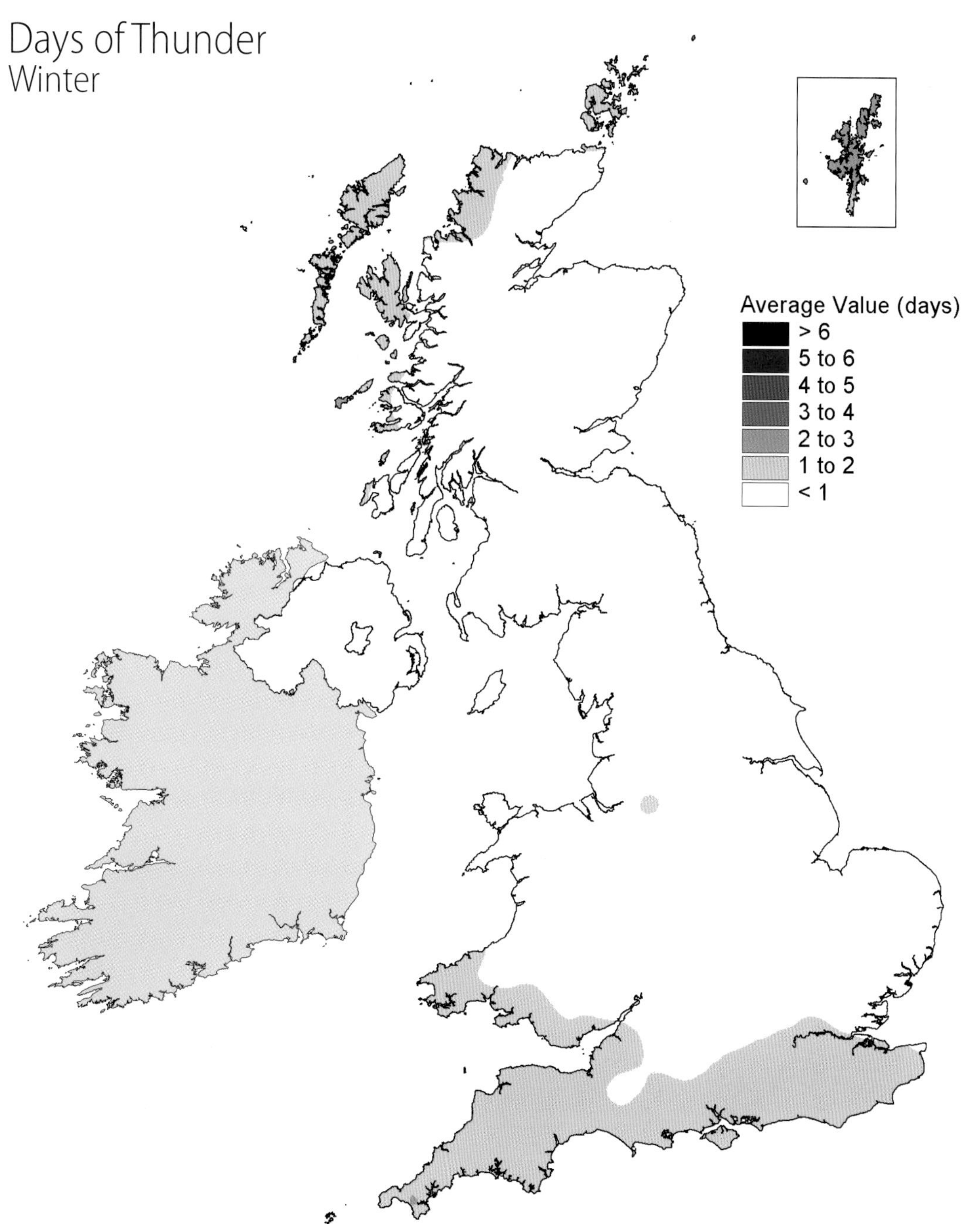

Average Value (days)
> 6
5 to 6
4 to 5
3 to 4
2 to 3
1 to 2
< 1

Days of Hail

by Season

These maps show the seasons and places prone to hail, with winter and spring and the northern and western fringes of Scotland seeing it most often. This usually falls from showers in cool northerly or westerly airstreams. Hailstones grow in size as they circulate through deep cloud, acquiring layers of ice. Most hailstones are about the size of a pea, but some grow larger and can damage cars and crops. In September 1958, hailstones the size of golf balls fell in West Sussex.

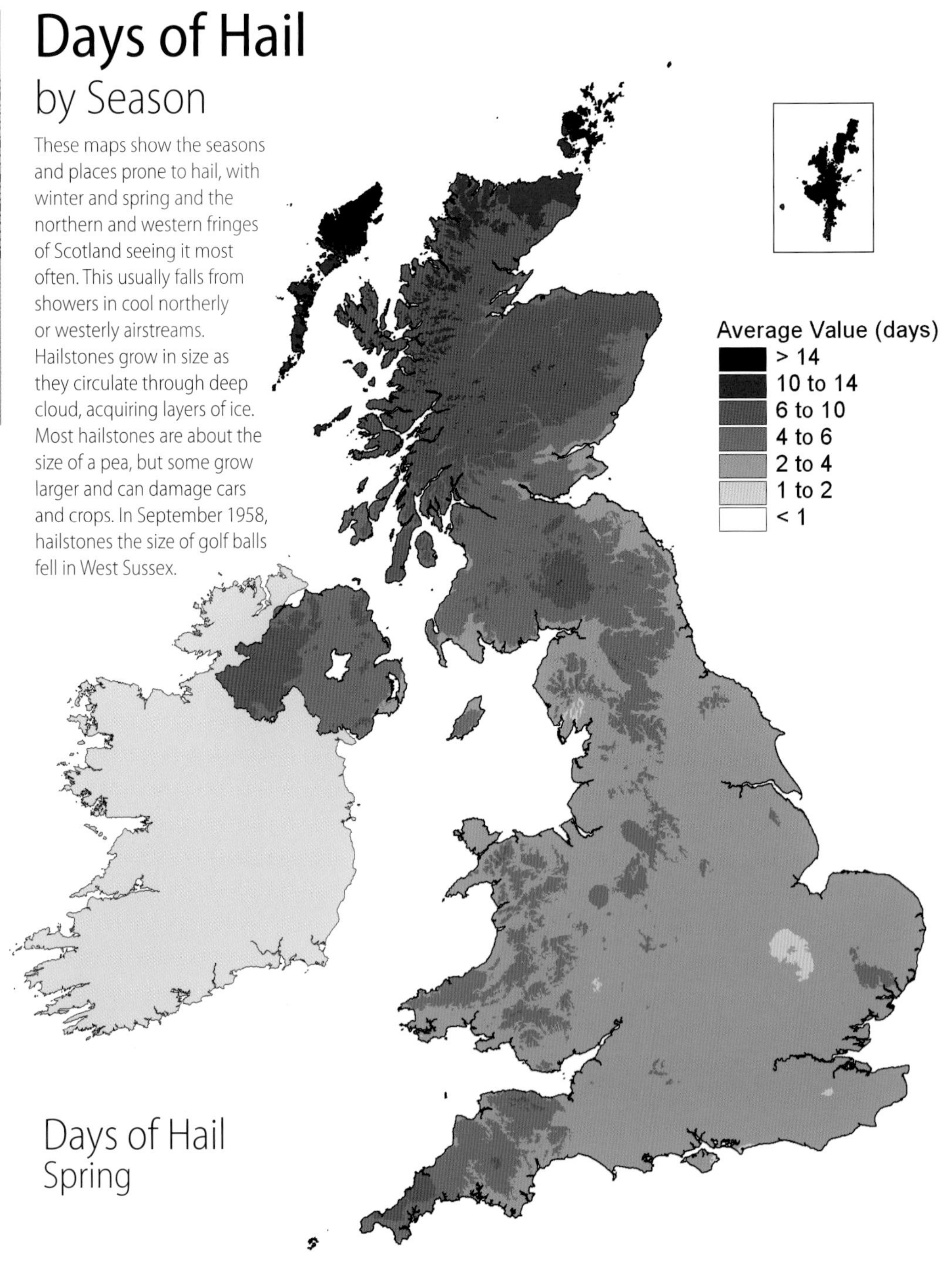
Average Value (days)
> 14
10 to 14
6 to 10
4 to 6
2 to 4
1 to 2
< 1

Days of Hail
Spring

Days of Hail
Summer

EXTREME WEATHER
BY SEASON

Average Value (days)

0.5 to 1.5

< 0.5

Days of Hail
Autumn

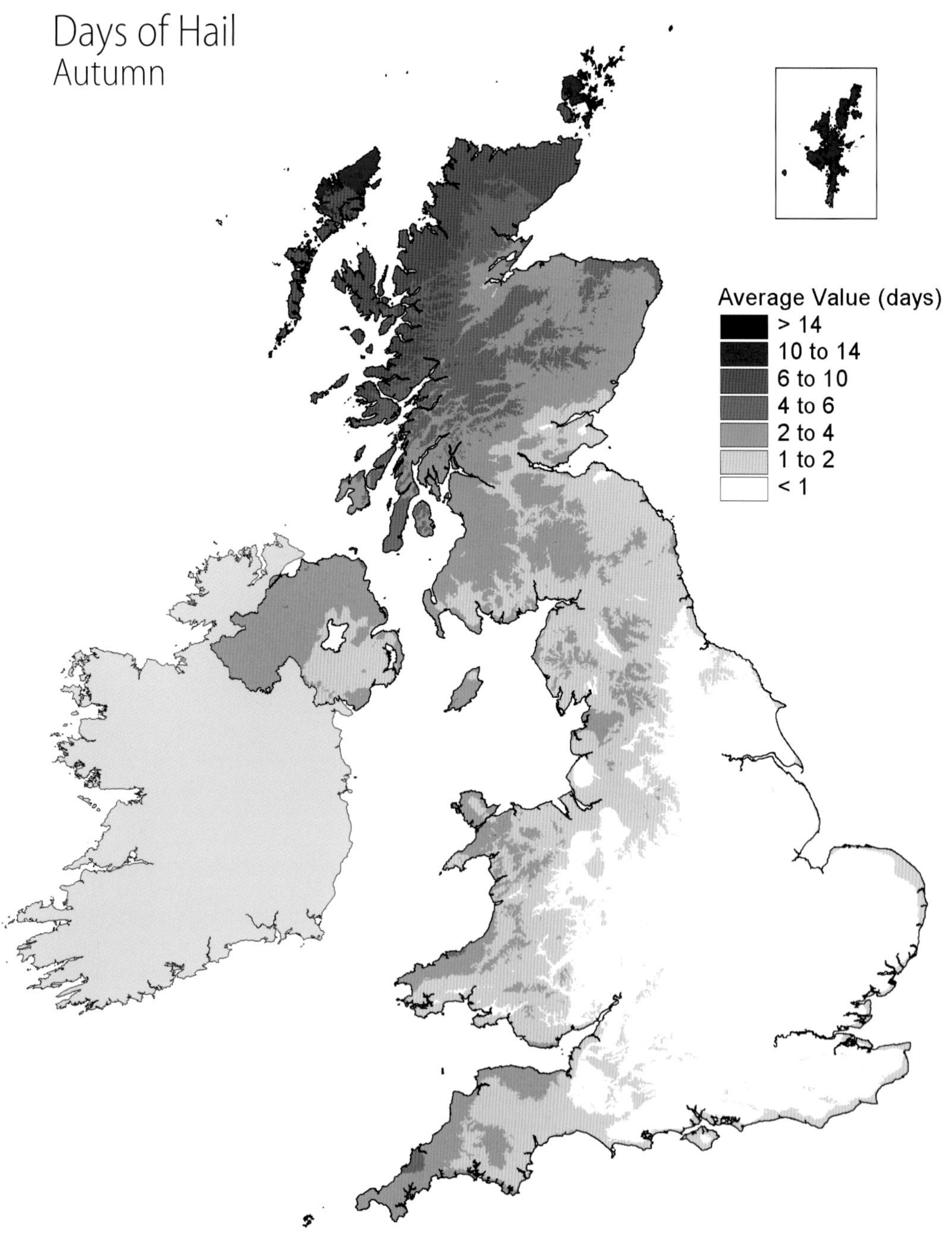
Average Value (days)
> 14
10 to 14
6 to 10
4 to 6
2 to 4
1 to 2
< 1

Days of Hail
Winter

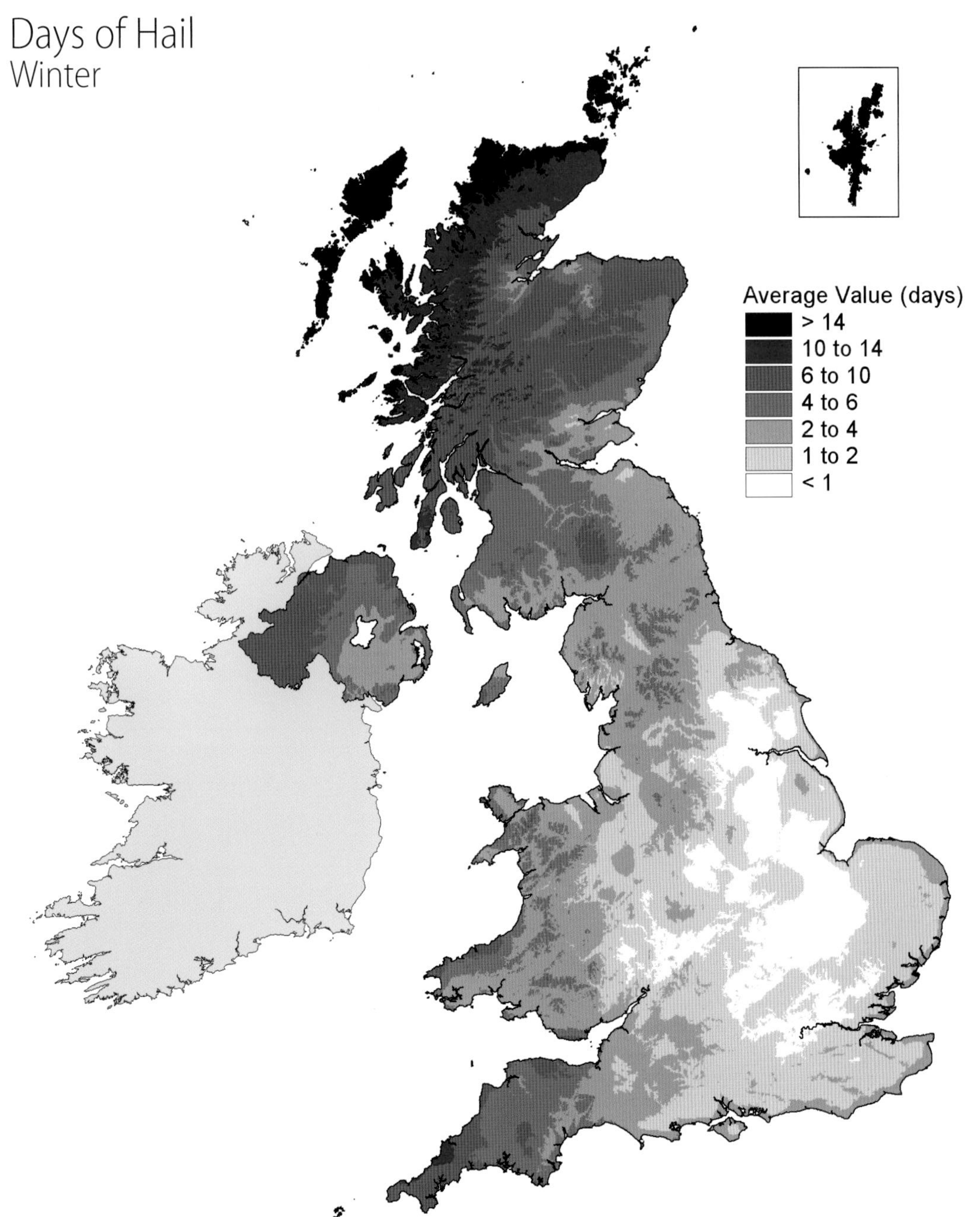
Average Value (days)
> 14
10 to 14
6 to 10
4 to 6
2 to 4
1 to 2
< 1

Annual Averages

These annual maps reflect the monthly and seasonal patterns described in earlier chapters. They also reveal which parts of the country are consistently the warmest, coldest, driest, wettest and so on.

In terms of sunshine, the favoured spots are on the south coast of England with over 1,850 hours a year on average, about 40% of the maximum number of hours that are possible. In contrast, Shetland manages less than 1,100 hours per year – only about a quarter of the hours possible.

The annual average-rainfall map tends to mirror a topographical map, with the wetter areas to the west and north. Wettest of all are the highest parts of Snowdonia and the Lake District where over 4,000 mm of rain falls in a typical year, with the western Scottish Highlands almost as wet. The driest areas are those near the Essex coast and the Thames Estuary, receiving less than 550 mm per year. However, this area, along with the rest of south-east England and East Anglia, is the most prone to thunderstorms.

In terms of mean temperature, the coldest areas are in Scotland's Grampian mountains with averages only a few degrees above freezing, over 100 frosts each year and more than 60 days of lying snow each year. The warmest places are central London and the west coast of Cornwall, but other southern coasts are not far behind. Here, frost and snow are relatively rare.

Upland and coastal areas are the most windy, especially those to the west and north such as the Western and Northern Isles of Scotland. Here we find the highest frequency of gales. The least windy places tend to be low-lying, well inland and towards the south, especially in valleys, forests and built-up areas.

Hours of Sunshine

Annual averages

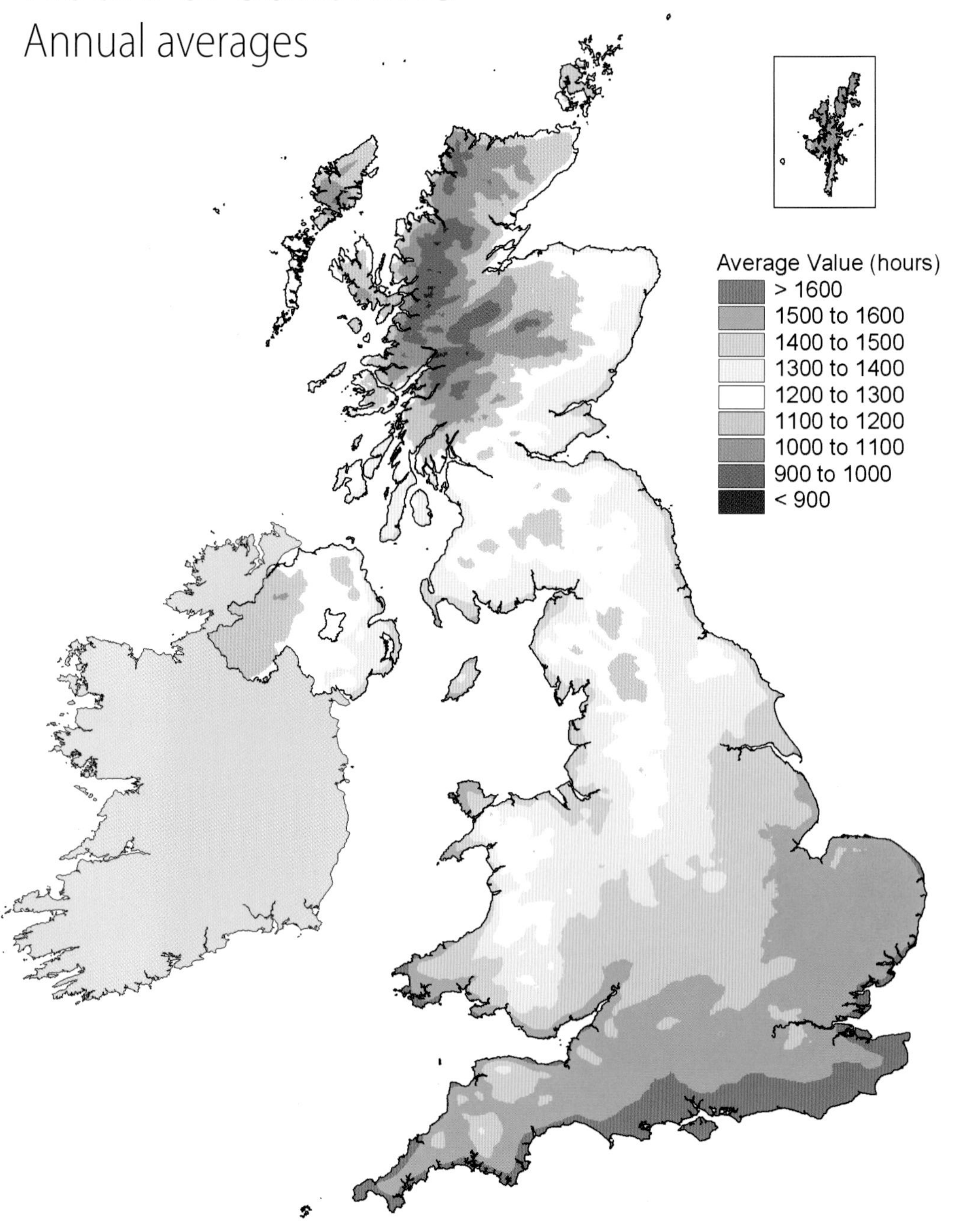
Average Value (hours)
> 1600
1500 to 1600
1400 to 1500
1300 to 1400
1200 to 1300
1100 to 1200
1000 to 1100
900 to 1000
< 900

Mean Temperature

Annual averages

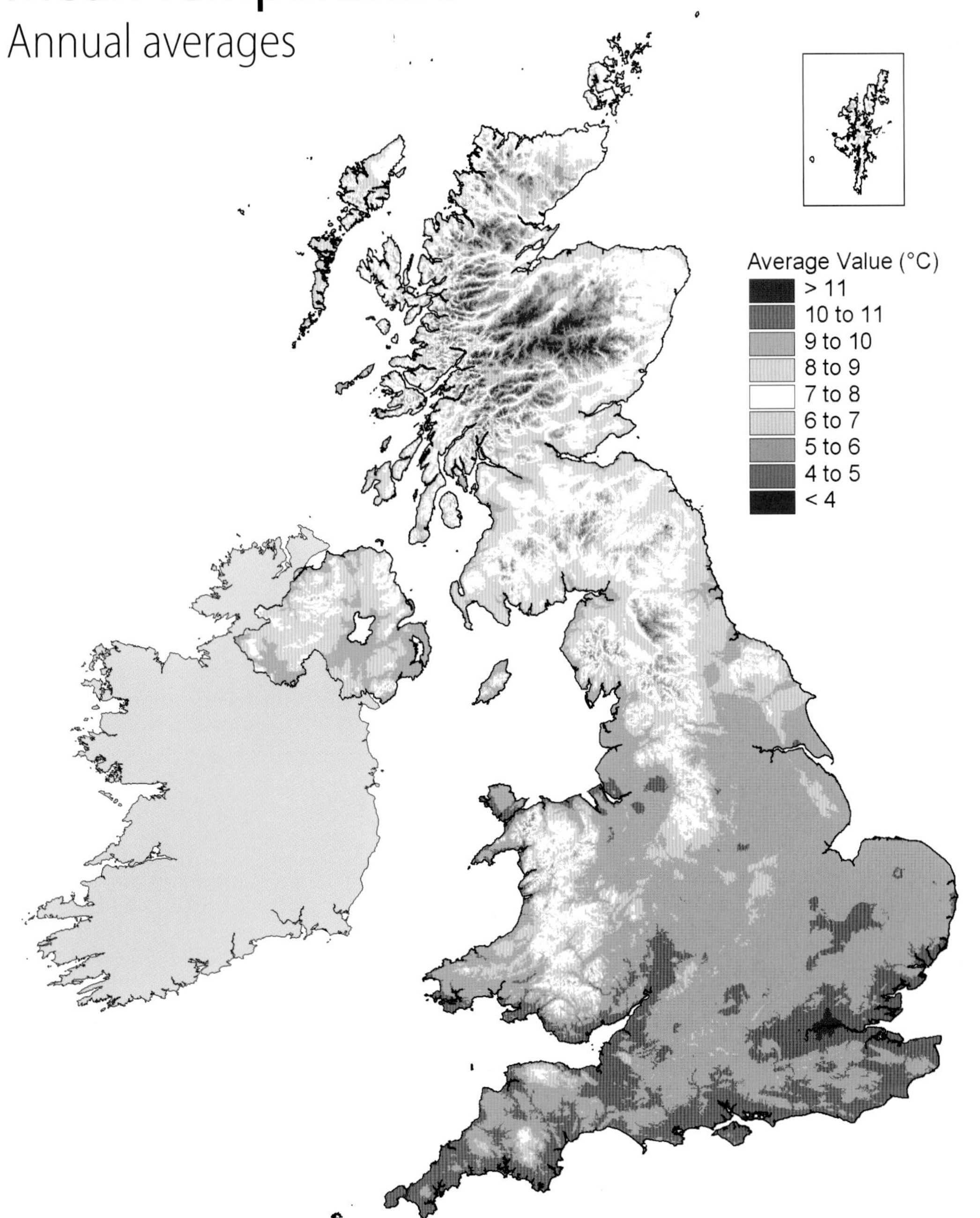
Average Value (°C)
> 11
10 to 11
9 to 10
8 to 9
7 to 8
6 to 7
5 to 6
4 to 5
< 4

Total Rainfall

Annual averages

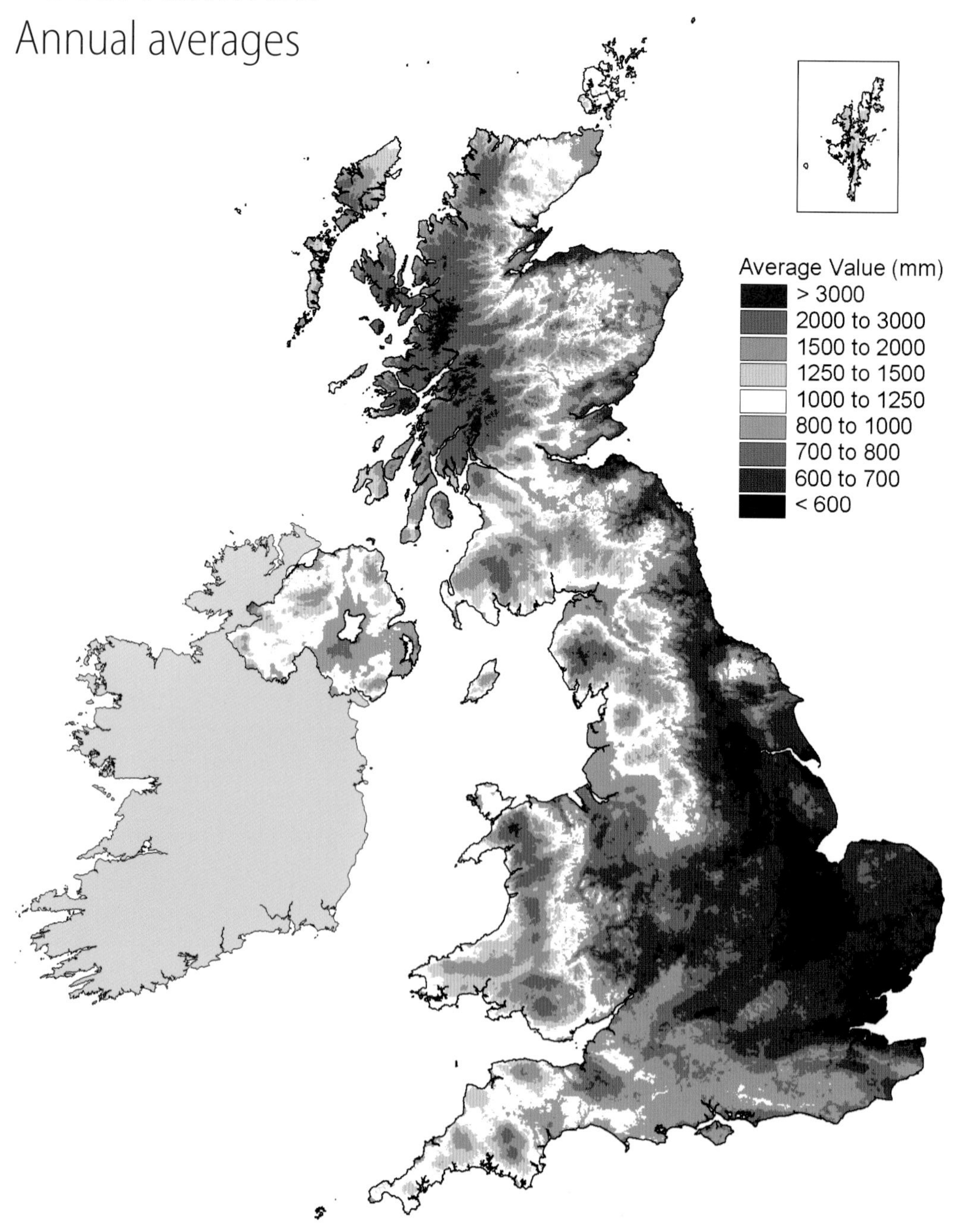

Average Value (mm)
> 3000
2000 to 3000
1500 to 2000
1250 to 1500
1000 to 1250
800 to 1000
700 to 800
600 to 700
< 600

Days of Rain 10mm or more

Annual averages

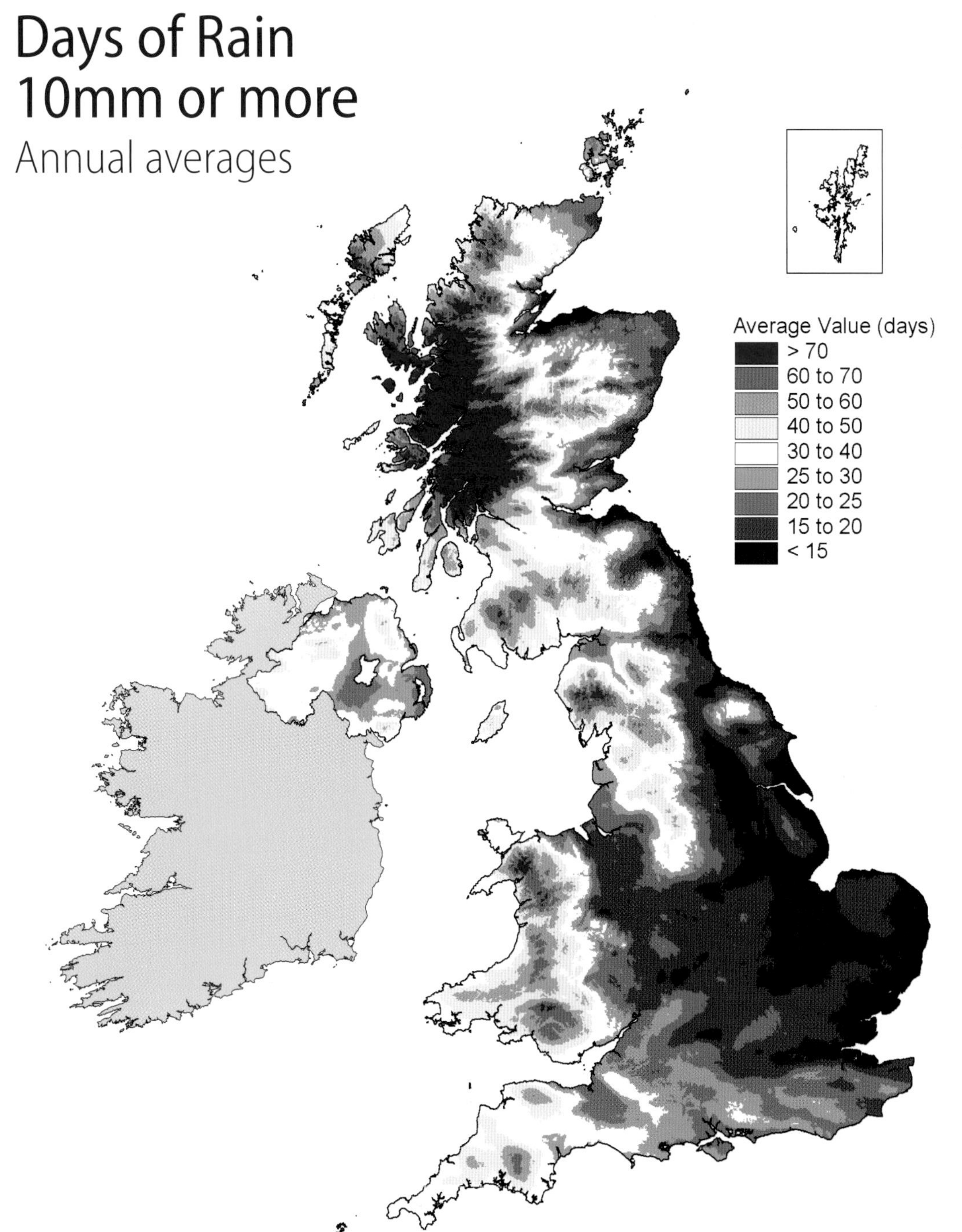
Average Value (days)
> 70
60 to 70
50 to 60
40 to 50
30 to 40
25 to 30
20 to 25
15 to 20
< 15

Wind Speed
Annual averages

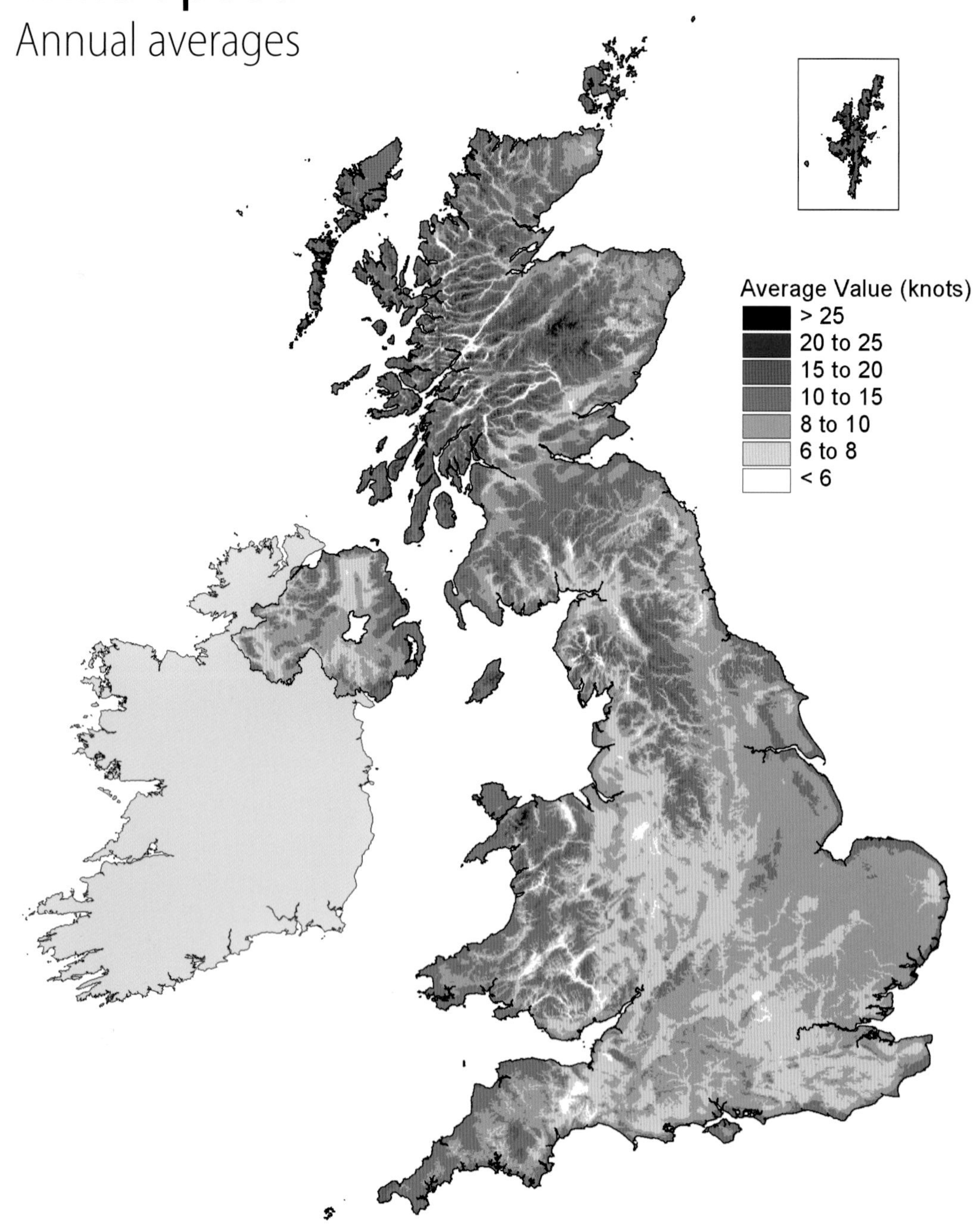
Average Value (knots)
> 25
20 to 25
15 to 20
10 to 15
8 to 10
6 to 8
< 6

Days of Ground Frost

Annual averages

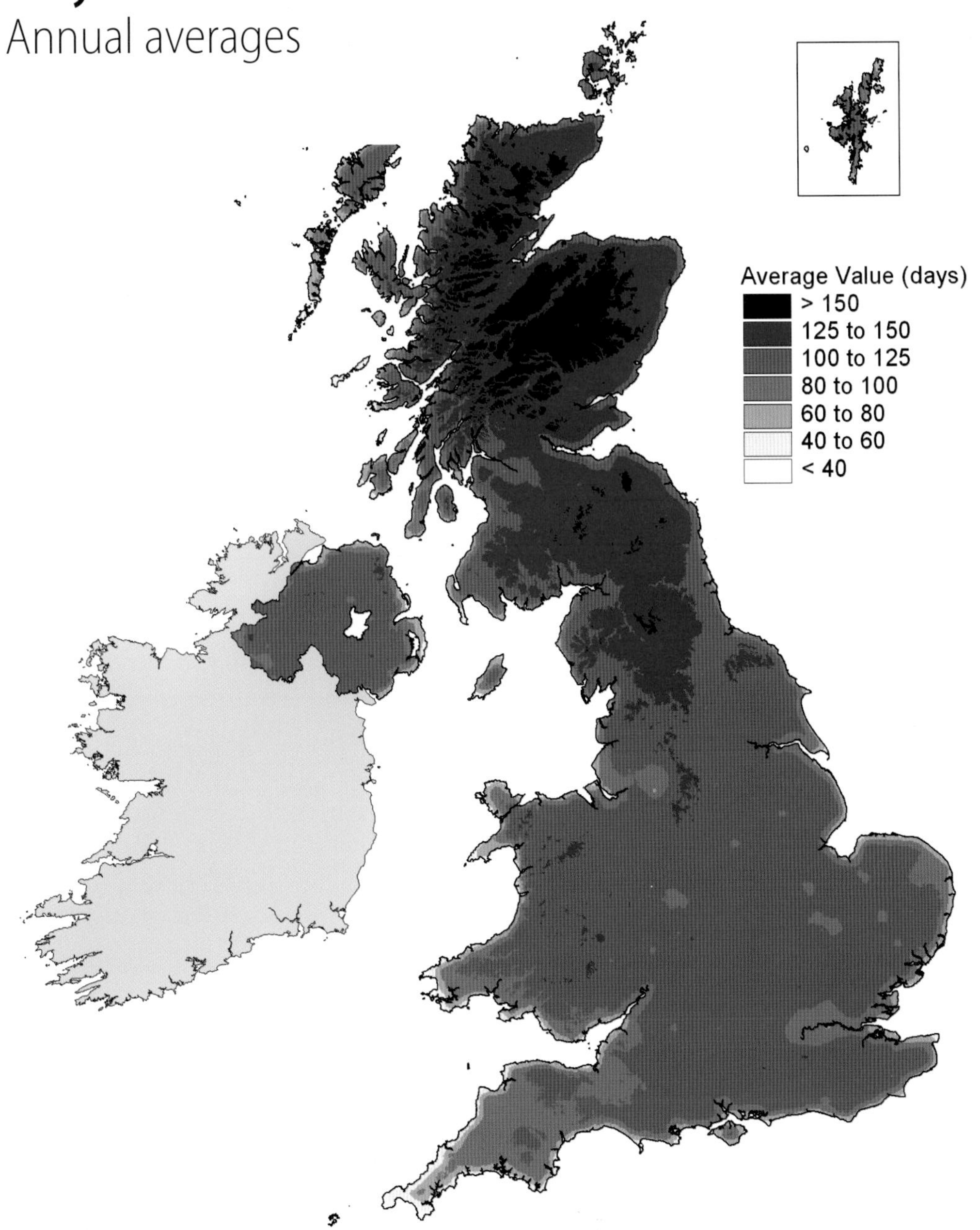

Average Value (days)
> 150
125 to 150
100 to 125
80 to 100
60 to 80
40 to 60
< 40

Days of Air Frost

Annual averages

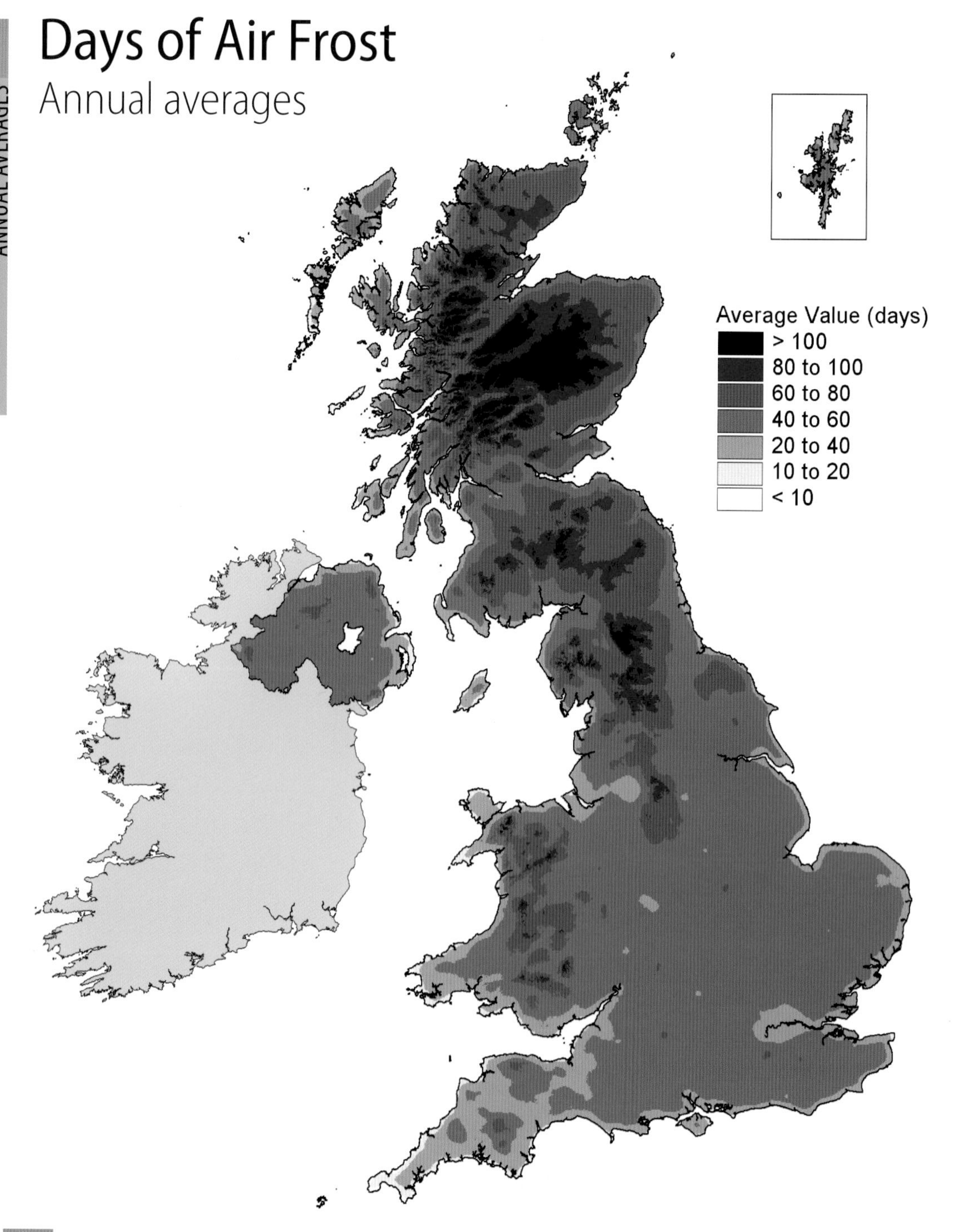
Average Value (days)
> 100
80 to 100
60 to 80
40 to 60
20 to 40
10 to 20
< 10

Days of Sleet/Snow Falling

Annual averages

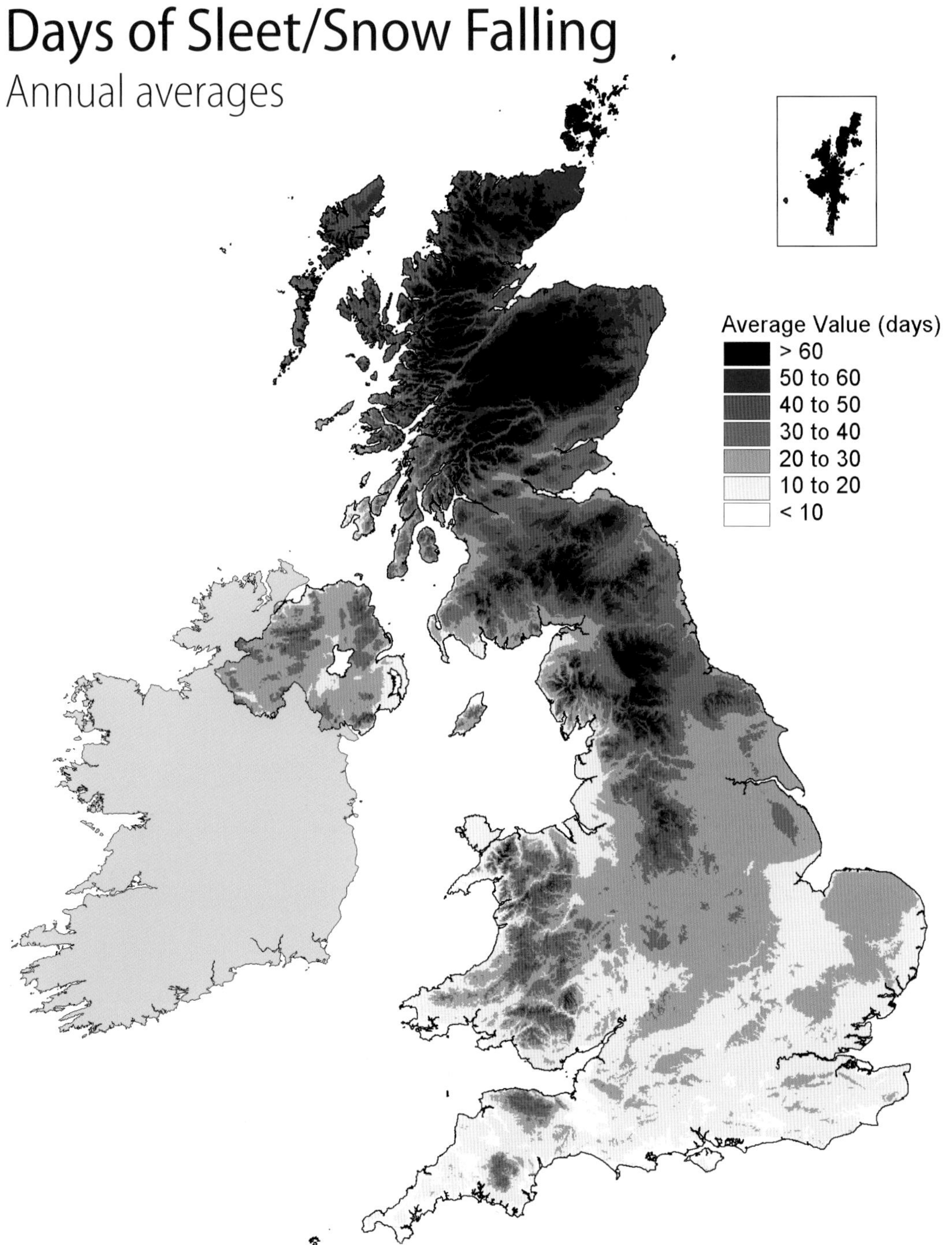
Average Value (days)
> 60
50 to 60
40 to 50
30 to 40
20 to 30
10 to 20
< 10

Days of Snow Lying

Annual averages

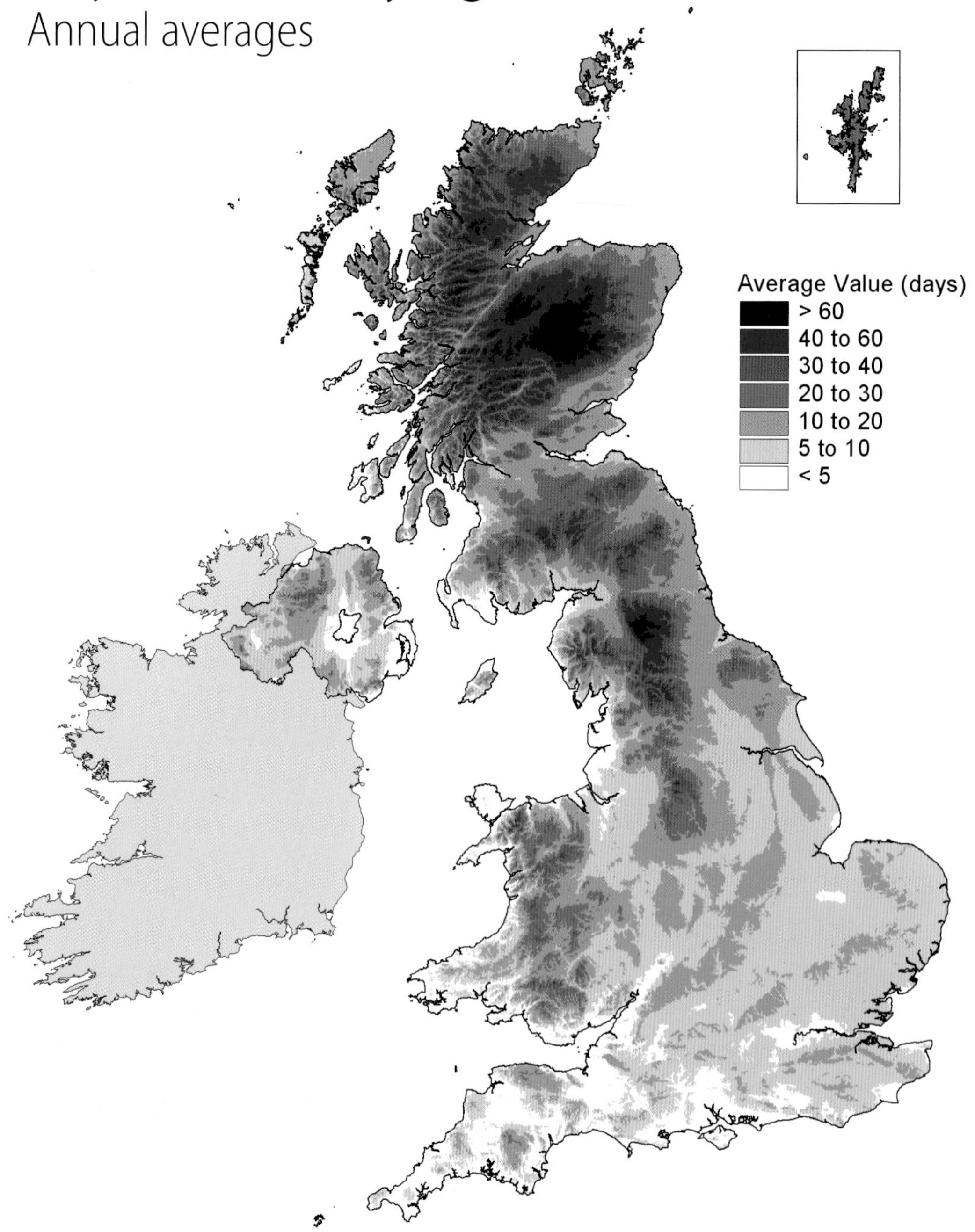
Average Value (days)
> 60
40 to 60
30 to 40
20 to 30
10 to 20
5 to 10
< 5

Days of Thunder

Annual averages

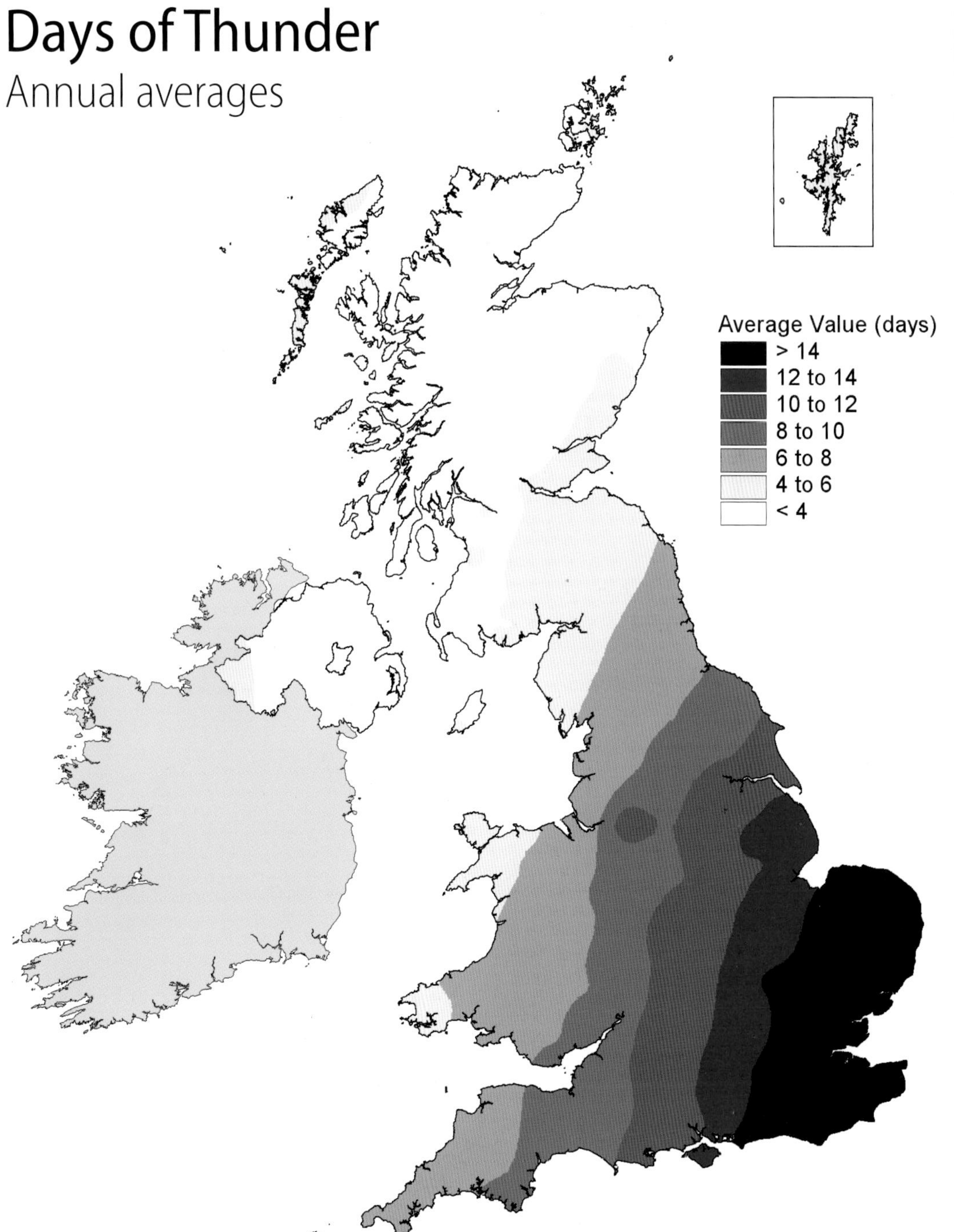
Average Value (days)
> 14
12 to 14
10 to 12
8 to 10
6 to 8
4 to 6
< 4

The Future of the British Weather

'All sorts of things and weather must be taken in together, to make up a year and a sphere.'
Ralph Waldo Emerson, 1803–1882

The climate is changing, globally and in the UK, and rarely does a day go by without a reference to this in the news. In 2007, the Intergovernmental Panel on Climate Change stated that it is very likely that man-made greenhouse gas* emissions have caused most of the observed increase in global average-temperatures since the mid-20th century. Recent international research has increased confidence in this statement. Our understanding of how the Earth's complex climate system works and our ability to simulate it in computer models have improved. These models have been used to predict Britain's possible future climate with the results presented in different scenarios of global greenhouse gas emissions, within different time periods and for different probabilities. A sample of these results is provided in the following maps.

There are three greenhouse gas emissions scenarios – low, medium and high – linked to different assumptions about the burning of fossil fuels. International efforts could reduce emissions significantly, but there is a time lag for the effect of gases that have already been emitted. The medium scenario assumes a balance between the use of fossil fuels and non-fossil energy sources, and that scenario is used here.

The time periods used are 2010 to 2039 (representing the 2020s), 2040 to 2069 (the 2050s) and 2070 to 2099 (the 2080s). Each map shows the expected changes, compared to the climate at the end of the 20th century.

Projections of climate change take into account uncertainties due to natural variability and the incomplete representation of the climate system in models. The probabilities used are 10%, 50% and 90%. The evidence equally points to change being above or below the 50% level, and this is referred to as the 'central estimate' of change. The 10% level represents changes that are very likely to be exceeded and 90% those very unlikely to be exceeded. For example, temperatures are expected to rise across the UK, with more warming in summer than in winter. Under the medium-emissions scenario, by the 2080s the rise in the summer average-temperature over southern England is very likely to be above 2°C but below 6.5°C, with a central estimate of 4°C.

Changes in the frequency of extreme weather, such as hot days or very wet days, are likely to have even more impact than changes in average conditions, so maps of projections for the warmest summer day and the wettest winter day are also included in this chapter.

* The greenhouse analogy arose because some atmospheric gases, principally carbon dioxide, water vapour and ozone, act like the glass in a greenhouse – allowing in sunlight but not letting out heat.

Summer Mean Temperature

Summer is the months of June, July and August. Central estimates of the average summer temperature increase through time, so that by the 2080s they are between 3°C and 4°C higher than now. All areas are expected to warm more in summer than in winter.

Emissions levels

There are three greenhouse gas emission scenarios – low, medium and high. These are linked to different assumptions about the burning of fossil fuels.

Medium

The medium scenario, used here, assumes a balance between the use of fossil fuels and and non-fossil energy sources.

Probability

Projections of change take into account uncertainties due to natural variability and the incomplete representation of the climate system in models.

10%

Changes are very likely to be exceeeded.

50%

Change may be above of below this central estimate.

90%

Changes are extremely unlikely to be exceeded.

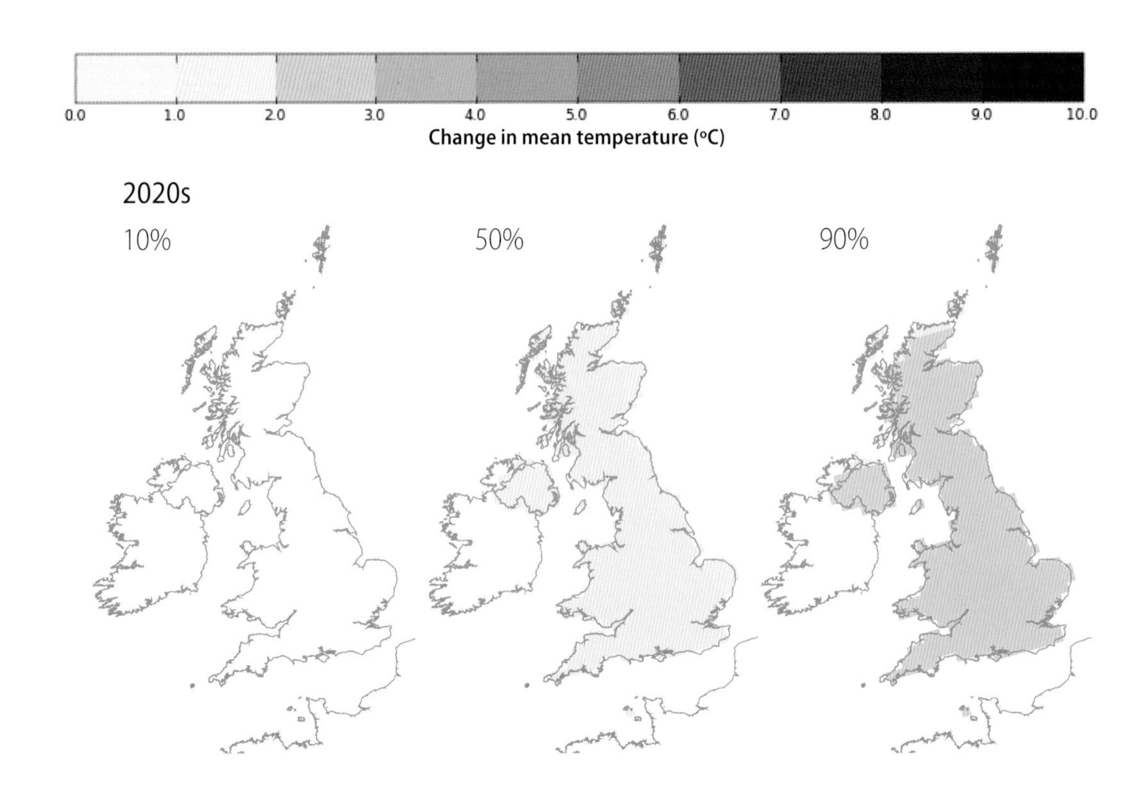
0.0
1.0
2.0
3.0
4.0
5.0
6.0
7.0
8.0
9.0
10.0
Change in mean temperature (°C)
2020s
10%
50%
90%

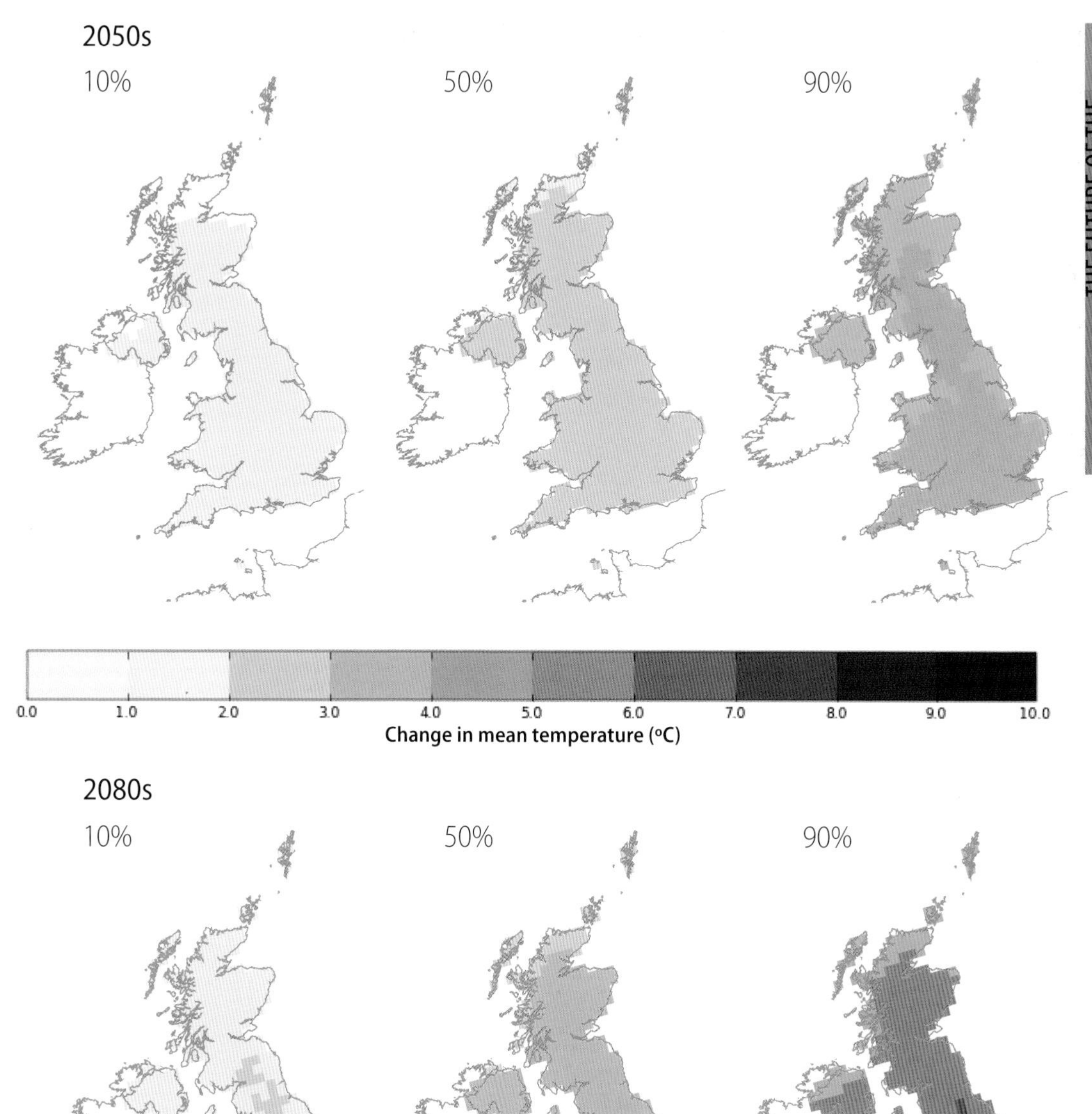
2050s
10%
50%
90%
0.0
1.0
2.0
3.0
4.0
5.0
6.0
7.0
8.0
9.0
10.0
Change in mean temperature (°C)
2080s
10%
50%
90%

Winter Mean Temperature

Winter is the months of November, December and January. Central estimates of the average winter temperature increase through time. By the 2080s they are generally between 2°C and 3°C higher than at present, but over 3°C higher in East Anglia and south-east England.

Emissions levels

There are three greenhouse gas emissions scenarios – low, medium and high. These are linked to different assumptions about the burning of fossil fuels.

Medium

The medium scenario, used here, assumes a balance between the use of fossil fuels and and non-fossil energy sources.

Probability

Projections of change take into account uncertainties due to natural variability and the incomplete representation of the climate system in models.

10%

Changes are very likely to be exceeeded.

50%

Change may be above of below this central estimate.

90%

Changes are unlikely to be exceeded.

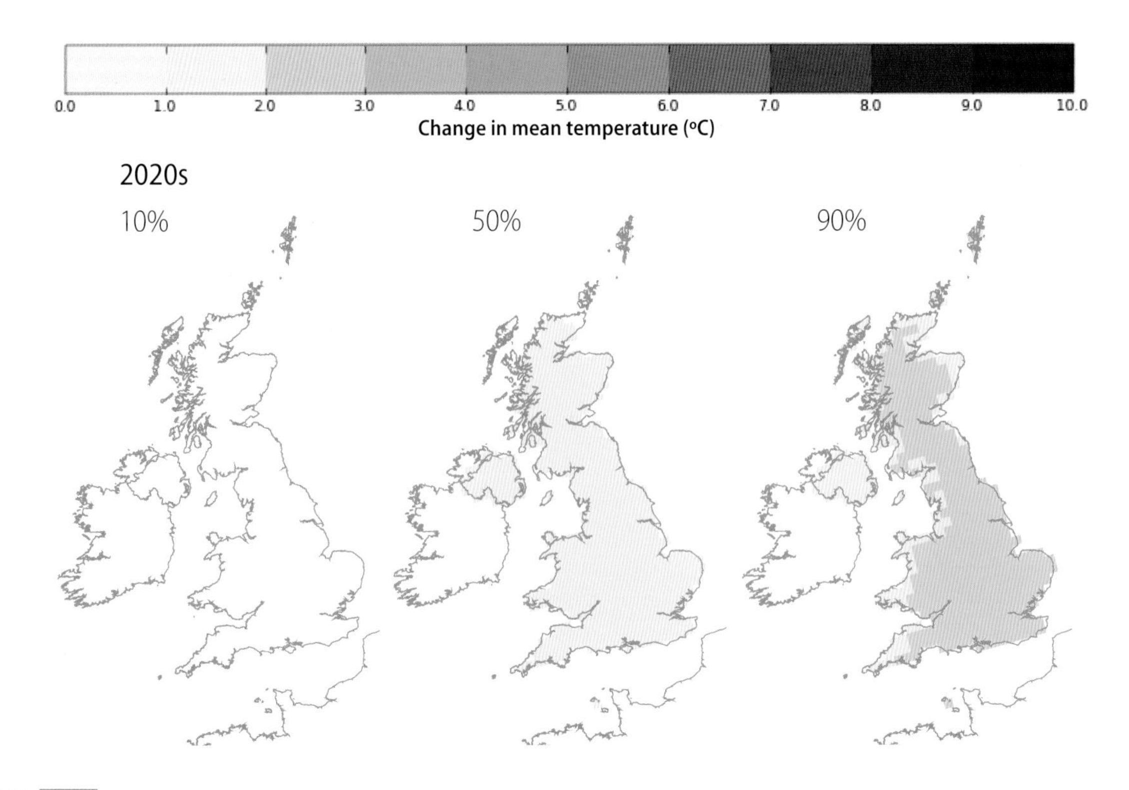
0.0
1.0
2.0
3.0
4.0
5.0
6.0
7.0
8.0
9.0
10.0
Change in mean temperature (°C)
2020s
10%
50%
90%

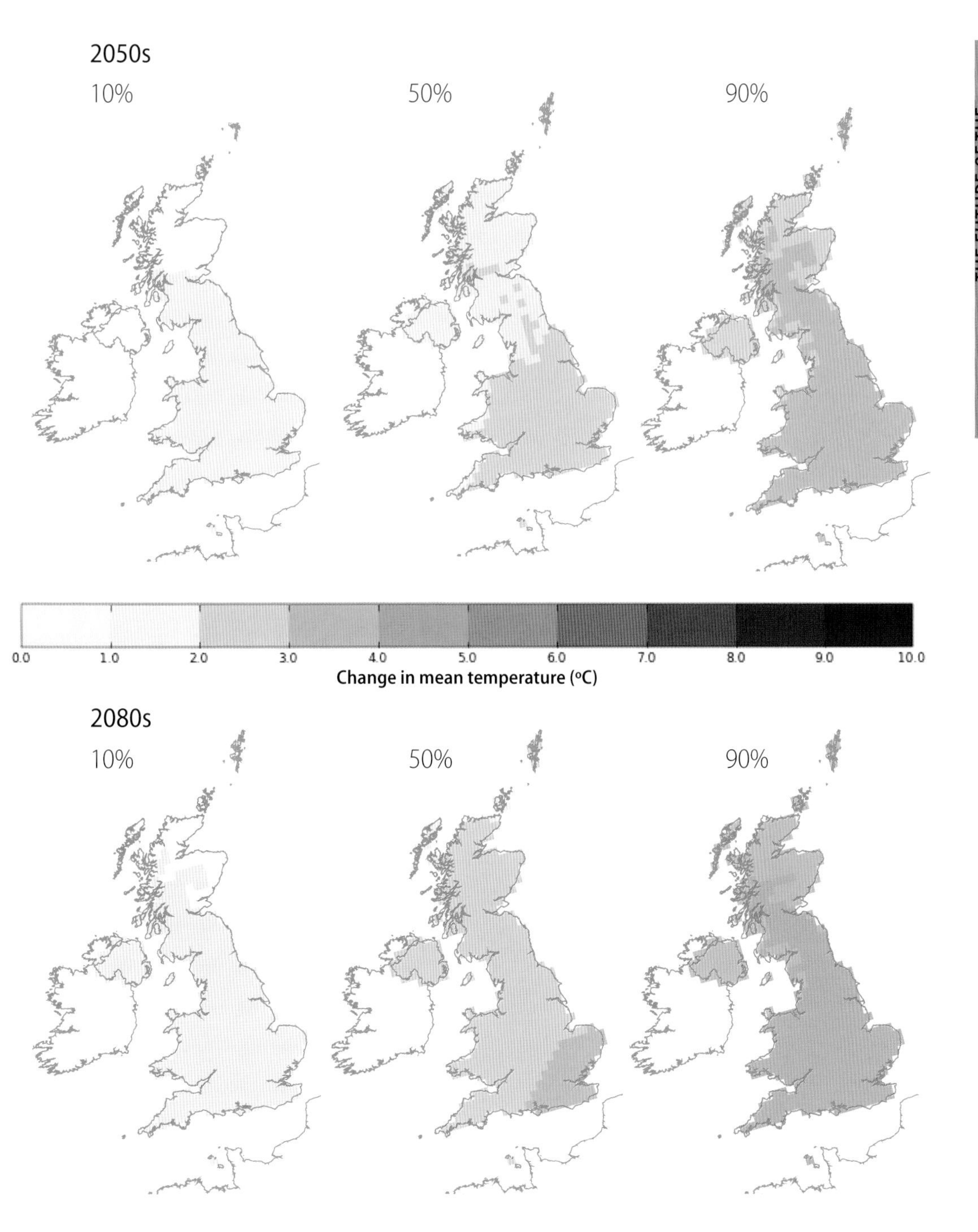

2050s
10%
50%
90%
0.0
1.0
2.0
3.0
4.0
5.0
6.0
7.0
8.0
9.0
10.0
Change in mean temperature (ºC)
2080s
10%
50%
90%

Warmest Day of Summer

In the current climate, the typical peak temperature on the warmest day each summer varies from 30°C in south-east England to 25°C in Northern Ireland and Scotland. Central estimates for the 2080s show the largest increases in the northern half of Britain, where peak temperatures will be about 5°C higher than now. However, increases as large as 10°C can't be ruled out (as shown in the 90% probability map). Temperatures may even go down (10% probability map).

Emissions levels

There are three greenhouse gas emissions scenarios – low, medium and high. These are linked to different assumptions about the burning of fossil fuels.

Medium

The medium scenario, used here, assumes a balance between the use of fossil fuels and and non-fossil energy sources.

Probability

Projections of change take into account uncertainties due to natural variability and the incomplete representation of the climate system in models.

10%

Changes are very likely to be exceeeded.

50%

Change may be above of below this central estimate.

90%

Changes are unlikely to be exceeded.

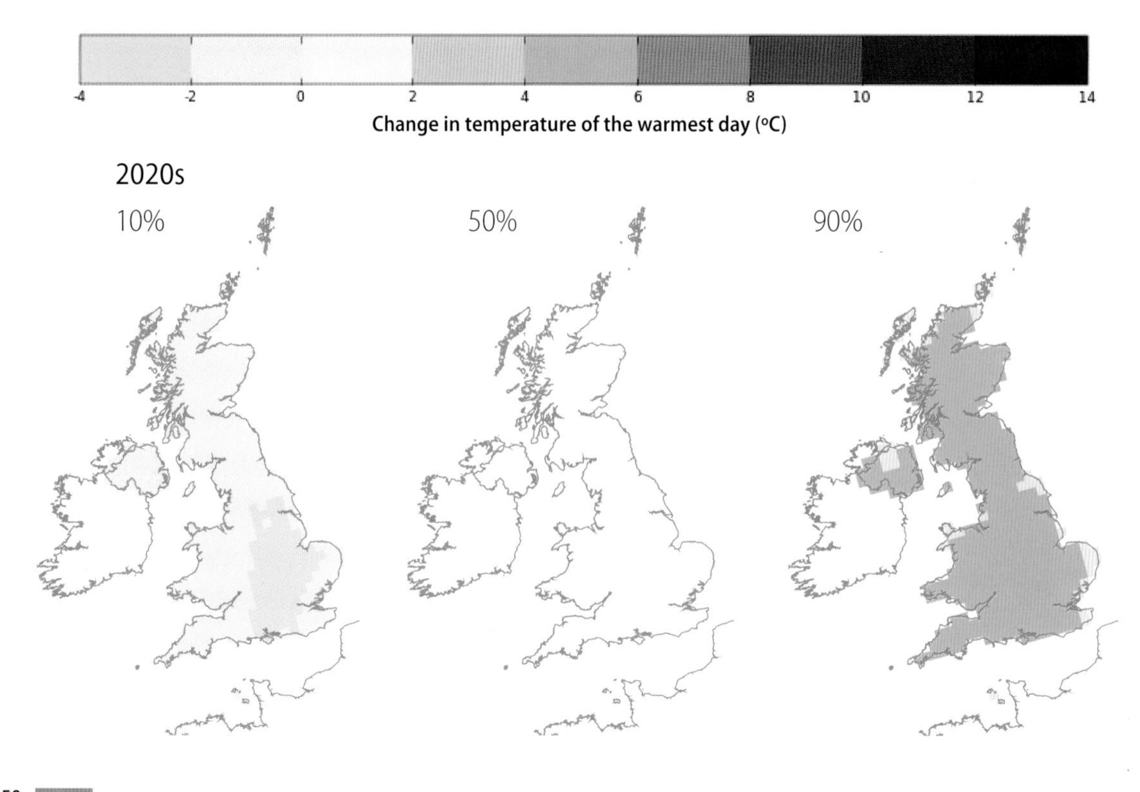
-4
-2
0
2
4
6
8
10
12
14
Change in temperature of the warmest day (°C)
2020s
10%
50%
90%

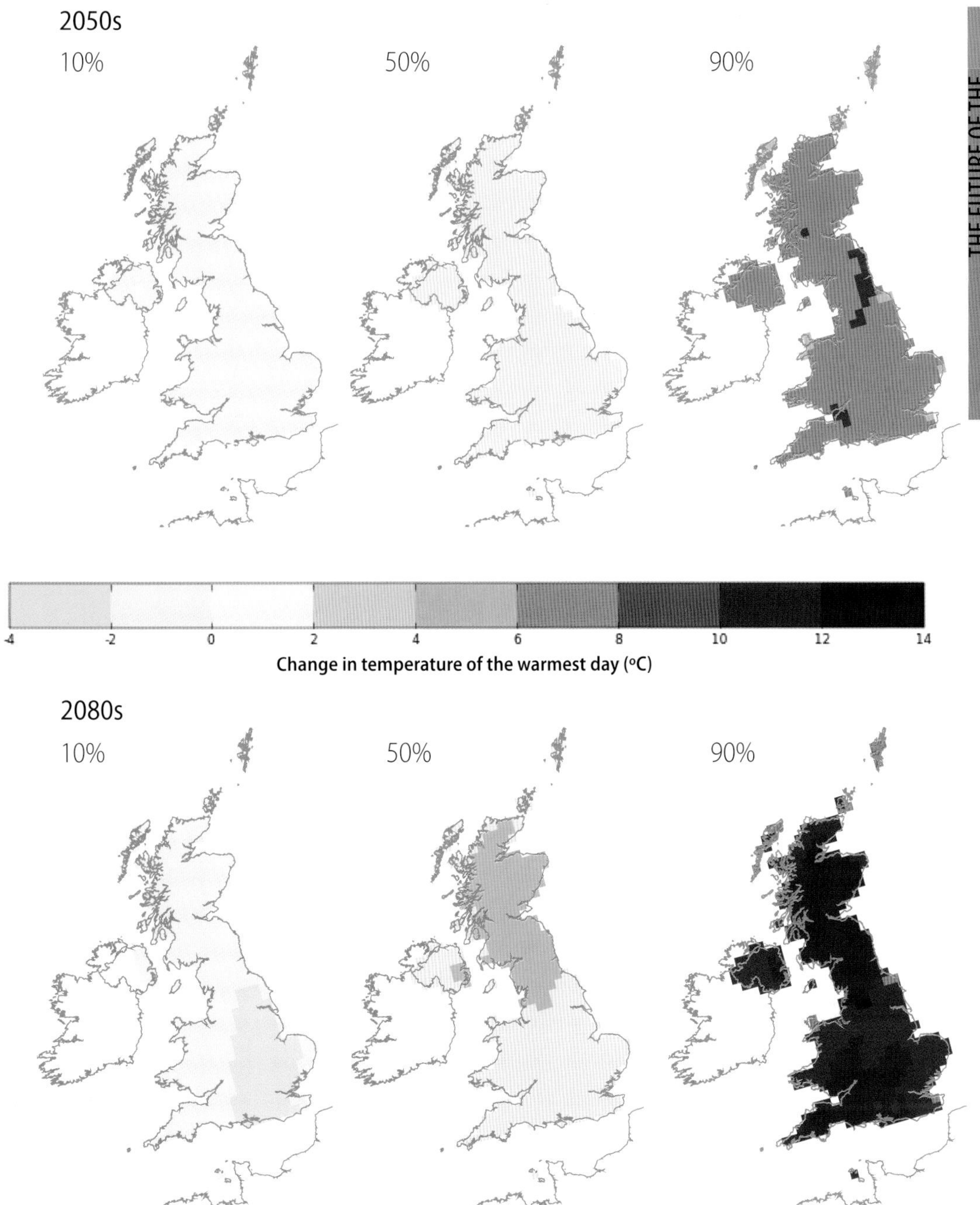
2050s
10%
50%
90%
-4
-2
0
2
4
6
8
10
12
14
Change in temperature of the warmest day (ºC)
2080s
10%
50%
90%

Wettest Day of Winter

In the current climate, the rainfall on the wettest day each winter is typically 28 mm in western Scotland, 23 mm in north-west England, 16 mm in the West Midlands and 12 mm in East Anglia. Central estimates show an upward trend with time so that by the 2080s the wettest days will generally record between 10% and 20% more rain.

Emissions levels

There are three greenhouse gas emissions scenarios – low, medium and high. These are linked to different assumptions about the burning of fossil fuels.

Medium

The medium scenario, used here, assumes a balance between the use of fossil fuels and and non-fossil energy sources.

Probability

Projections of change take into account uncertainties due to natural variability and the incomplete representation of the climate system in models.

10%

Changes are very likely to be exceeeded.

50%

Change may be above of below this central estimate.

90%

Changes are unlikely to be exceeded.

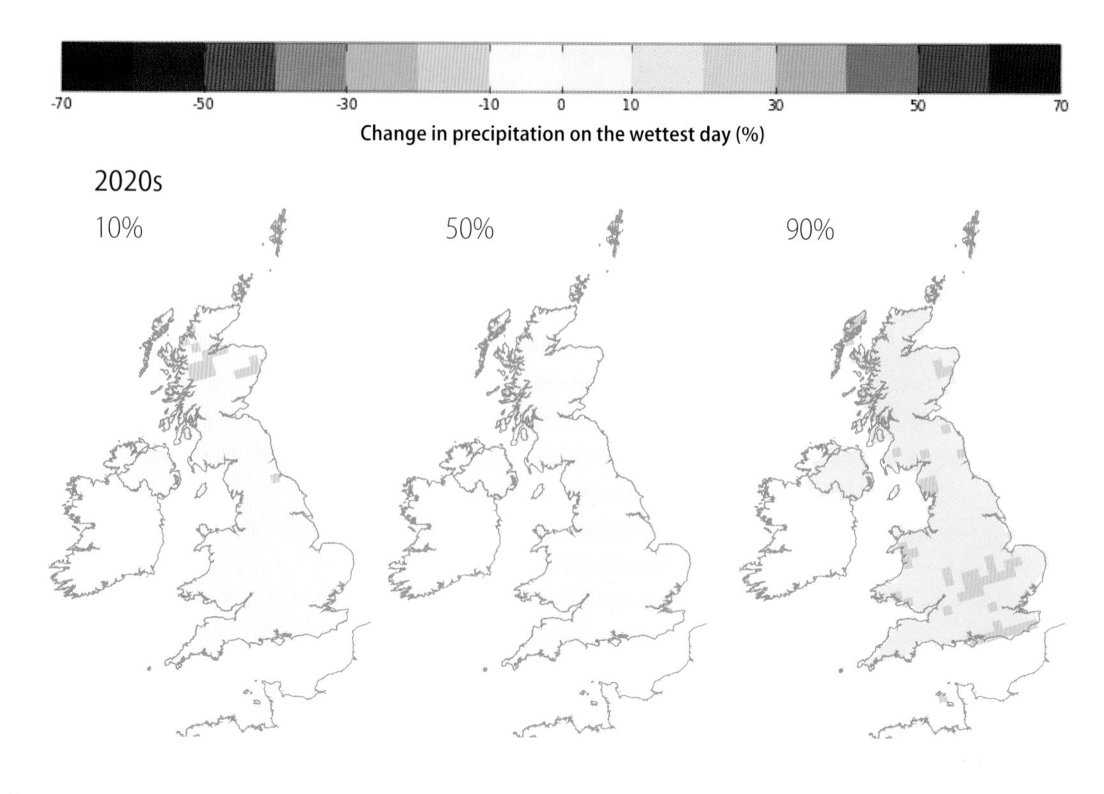
-70
-50
-30
-10
0
10
30
50
70
Change in precipitation on the wettest day (%)
2020s
10%
50%
90%

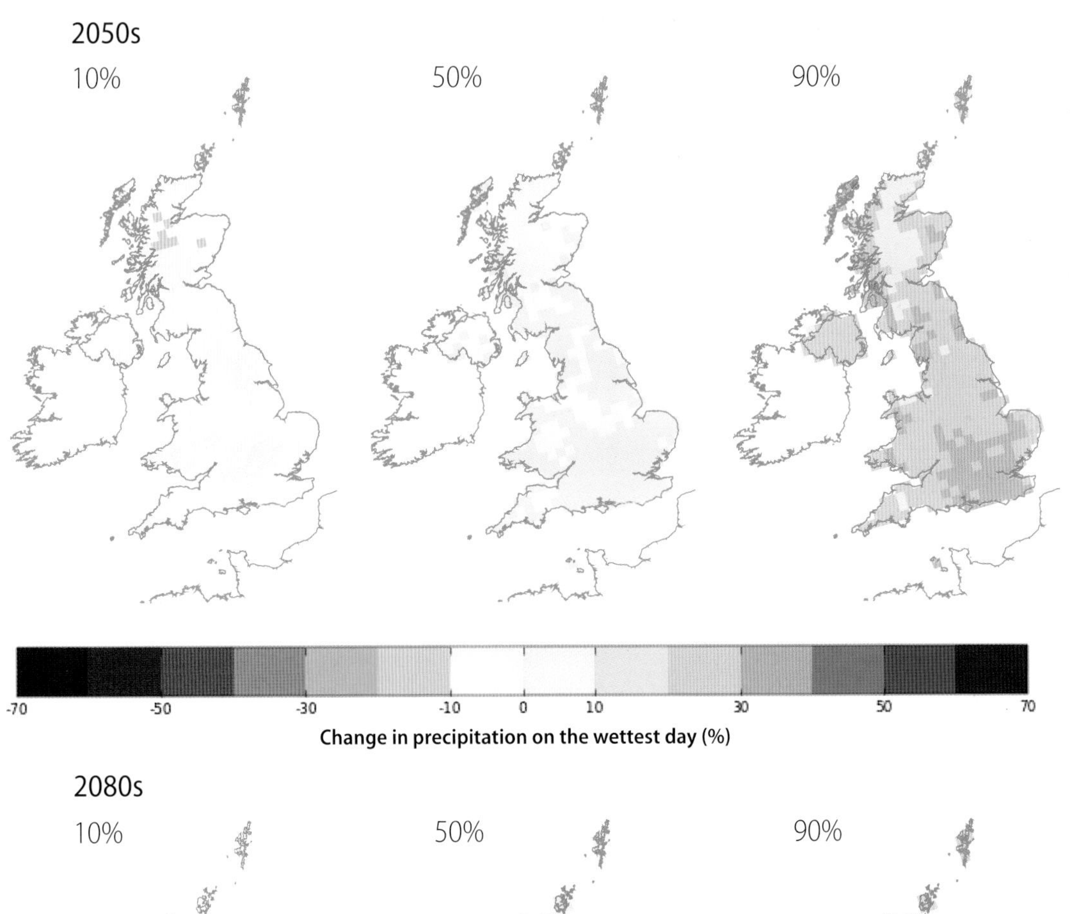
2050s
10%
50%
90%
-70
-50
-30
-10
0
10
30
50
70
Change in precipitation on the wettest day (%)

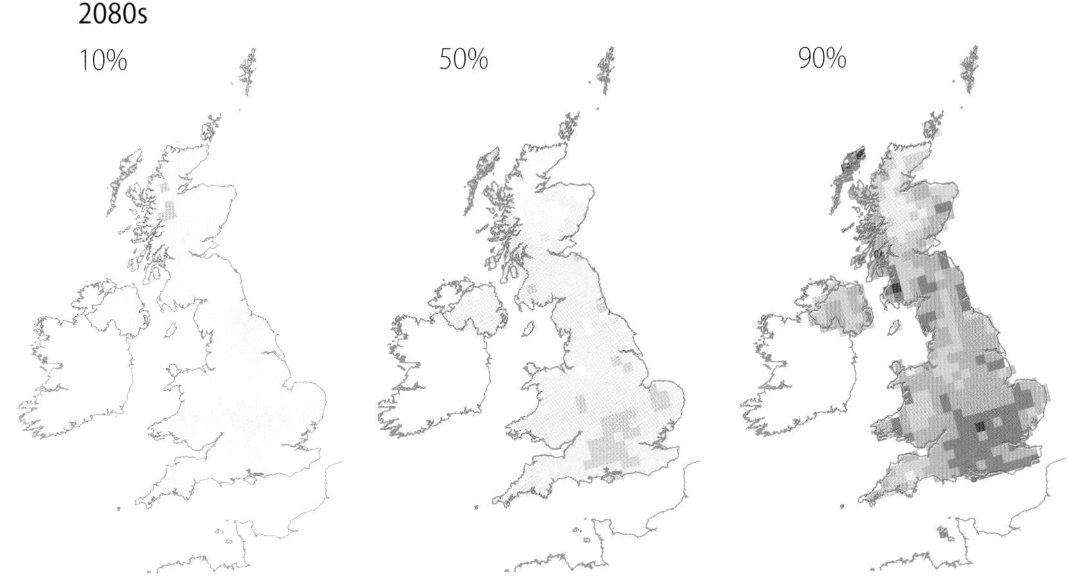
2080s
10%
50%
90%

Summer Mean Rainfall

Summer is the months of June, July and August. Summer rainfall is expected to decrease across the UK with time. Central estimates for the 2080s show decreases varying between 10% in the far north of Scotland and 30% in the southernmost counties of England.

Emissions levels

There are three greenhouse gas emissions scenarios – low, medium and high. These are linked to different assumptions about the burning of fossil fuels.

Medium

The medium scenario, used here, assumes a balance between the use of fossil fuels and and non-fossil energy sources.

Probability

Projections of change take into account uncertainties due to natural variability and the incomplete representation of the climate system in models.

10%

Changes are very likely to be exceeeded.

50%

Change may be above of below this central estimate.

90%

Changes are unlikely to be exceeded.

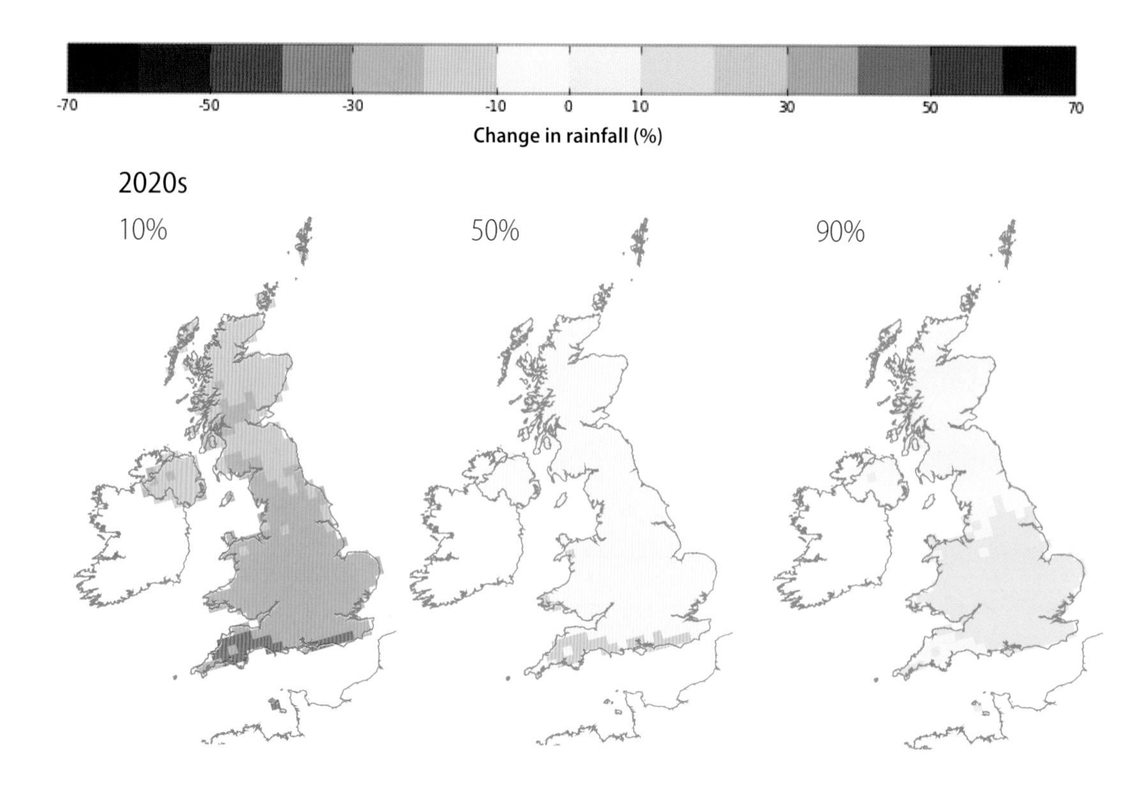
-70
-50
-30
-10
0
10
30
50
70
Change in rainfall (%)
2020s
10%
50%
90%

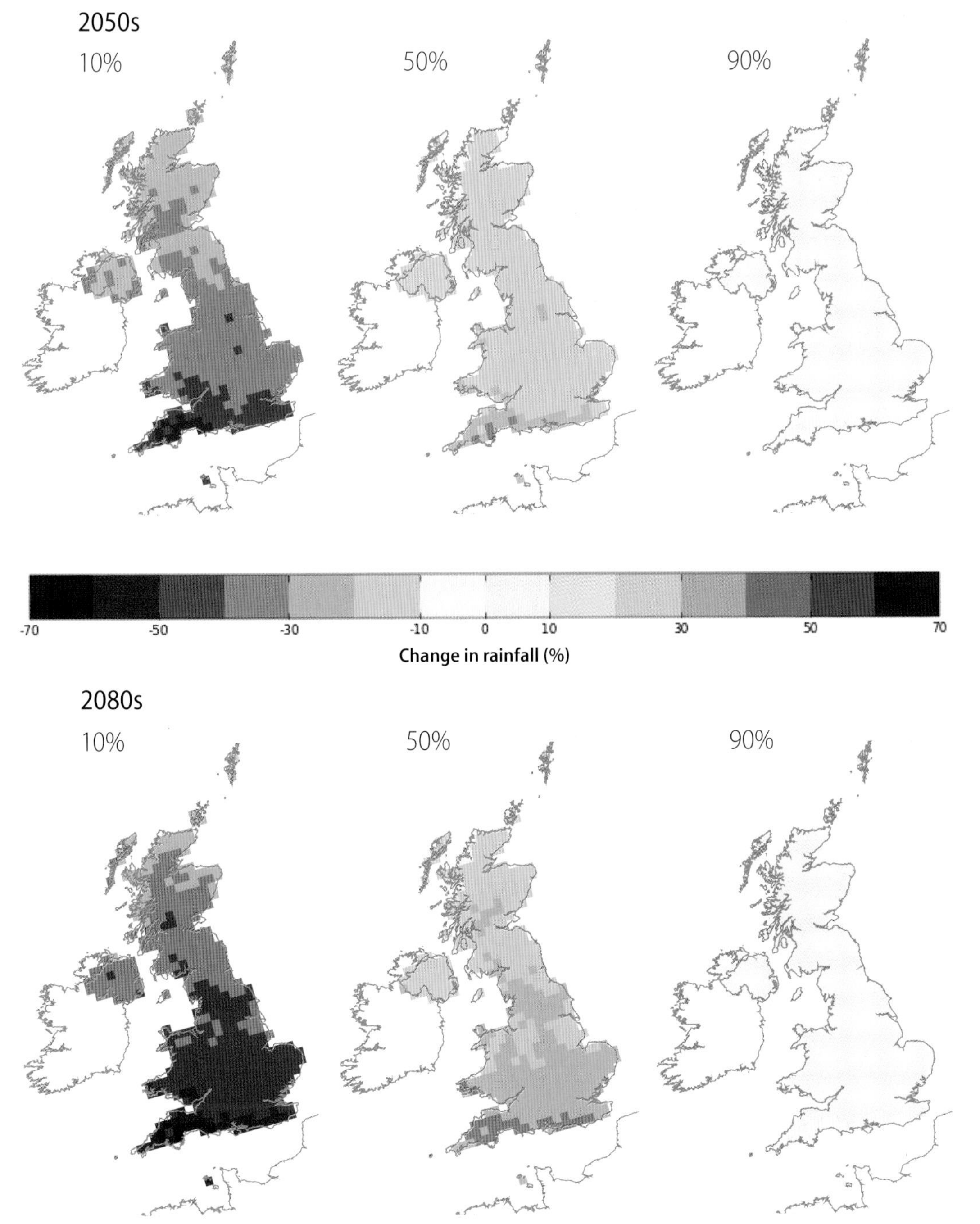
2050s
10%
50%
90%
-70
-50
-30
-10
0
10
30
50
70
Change in rainfall (%)
2080s
10%
50%
90%

Winter Mean Rainfall

Winter is the months of November, December and January and winter rainfall is projected to increase with time. Central estimates for the 2080s show increases between 10% and 20%, with slightly higher increases across much of southern England and in some coastal regions of the UK. Very little change is predicted across the Scottish Highlands.

Emissions levels

There are three greenhouse gas emissions scenarios – low, medium and high. These are linked to different assumptions about the burning of fossil fuels.

Medium

The medium scenario, used here, assumes a balance between the use of fossil fuels and and non-fossil energy sources.

Probability

Projections of change take into account uncertainties due to natural variability and the incomplete representation of the climate system in models.

10%

Changes are very likely to be exceeeded.

50%

Change may be above of below this central estimate.

90%

Changes are unlikely to be exceeded.

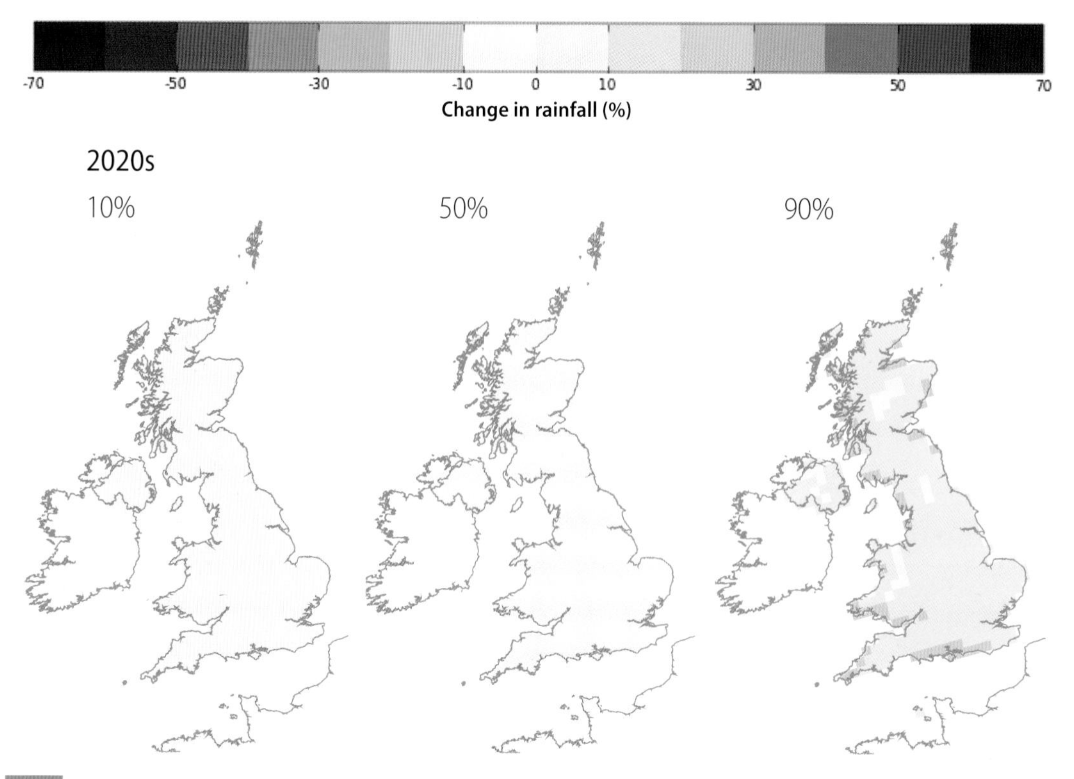
-70
-50
-30
-10
0
10
30
50
70
Change in rainfall (%)
2020s
10%
50%
90%

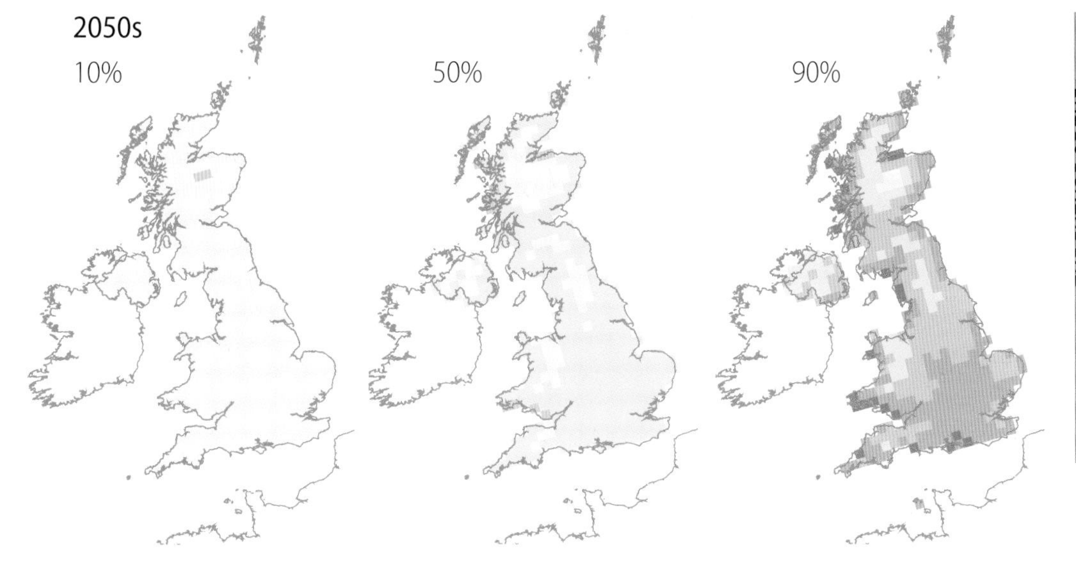
2050s
10%
50%
90%

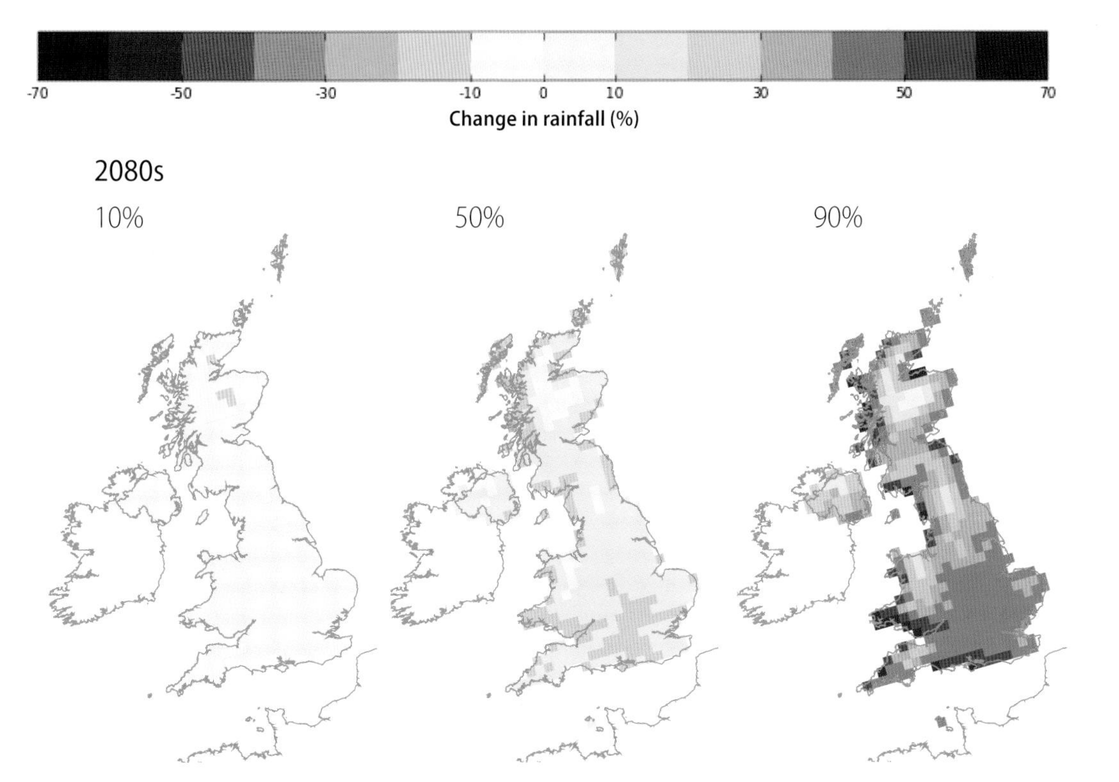
-70
-50
-30
-10
0
10
30
50
70
Change in rainfall (%)
2080s
10%
50%
90%

UK Climate Projections

The Earth's climate has changed many times in response to natural causes. The term 'climate change' usually refers to man-made changes that have occurred in the last 100 years. Human activities such as burning coal, oil and gas have led to an increase in atmospheric 'greenhouse' gases, such as carbon dioxide. As a result, there has been an increase in average temperatures, which is continuing. Globally, the temperature has risen by about 0.8°C since the late 19th century, with the decade 2000–2009 the warmest in the 160-year record. Even temperature changes that small could have serious consequences around the world, particularly for health, agriculture and the natural environment.

To help monitor changes in Britain, we can use the Central England Temperature (CET) series from 1659. For recent years this has been calculated from observations made at weather stations in Hertfordshire, Worcestershire and Lancashire. Long-term changes in the CET represent those across most of the country. After a period of relative stability for most of the 20th century, the CET has increased by about 1°C since the 1970s. The warmest year in this 350-year series was 2006 with several other recent years and their seasons among the top ten warmest. Studies have shown that this observed rate of warming cannot be explained by natural climate variability alone.

The UK Climate Projections released in 2009 (UKCP09) provide information on how Britain's climate is likely to change in the 21st century, as it responds to rising levels of greenhouse gases. The projections were produced from the results of multiple runs of a global climate model developed by the Met Office Hadley Centre and runs of 12 of the world's other leading climate models.

UKCP09 is a climate analysis tool that features the most comprehensive climate projections ever produced, explaining the likelihood of different levels of climate change in terms of the potential ranges. Uncertainty in projections arises from natural climate variability, incomplete representation of the Earth's climate in models and uncertainty in future man-made emissions of greenhouse gases and other pollutants.

Changes predicted by the climate models are likely to result in significant impacts and the UKCP09 projections are made available through the UK Climate Impacts Programme (UKCIP). This is a government-funded organisation that helps policy- and decision-makers in the public and private sectors to understand and plan for the effects of climate change, aiming to minimise any negative impacts and make the most of any positive ones.

We already have plenty of examples of the impact of extreme weather in Britain. Intense heat waves, such as that of 2003, can put a strain on the health, fire and transport services. Floods, such as those of summer 2007, can cause millions of pounds worth of damage to homes, businesses and the road network as well as interrupting water and power supplies.

Everyone in Britain will notice changes. In winter, we may not need to heat our homes and offices so much but more air conditioning might be required in summer. Gardeners and growers will benefit from a longer growing season and the opportunity to cultivate new crops, but the increased risk of summer heat waves and winter floods will not be so welcome.

We'll see changes in nature too – to the times that trees are in leaf and plants flower and fruit, to the migration of birds and insects and to the marine life off our shoreline. The winners and losers in a changing climate are yet to be discovered, but one thing we can be sure of – throughout the 21st century, the British will still be talking about the vagaries of the weather!

For further information about UKCIP see http://www.ukcip.org.uk/ and about UKCP09 see http://ukclimateprojections.defra.gov.uk/

Met Office weather data

The maps for 1971 to 2000 presented in this book rely upon a long series of regular weather observations made all over the country, to the same high standard. These observations are used for a variety of purposes, including the starting point of the numerical models that produce local to global weather forecasts; to answer questions about recent weather; for research purposes and to build up a picture of the climate.

Those places making observations primarily for weather forecasts and warnings are known as 'synoptic' stations. In order to meet other requirements, there is a further network of voluntary climate stations maintained by members of the public, schools, research stations and local authorities to name but a few. In addition there is a much larger network of 'rainfall stations' where only rainfall measurements are made regularly. In the UK at present there are about 200 synoptic, 400 climate and 2,700 rainfall stations.

Up until about 15 years ago, the vast majority of these observations were made by human observers. Gradually, observing stations are being automated so that almost all synoptic data is now collected from automatic weather stations. These have the advantage of collecting data in remote areas and at times when a human observer may not be available (e.g. at night-time or weekends). Although automation is underway, many climate and rainfall stations still rely upon a human observer taking readings at 0900 GMT each morning.

Whether automated or not, the appearance of a weather station is much the same. An instrument enclosure is set up on generally level ground and away from the immediate influence of obstructions such as buildings, fences or trees. It always contains a white, louvred thermometer screen, built to the 1864 specifications of British engineer Sir Thomas Stevenson, so that air temperatures can be measured at a height of 1.2 m above the ground. The Stevenson screen is accompanied by a 127 mm (5 inch) diameter rain gauge made of copper or stainless steel, sited so that the rim is 30 cm (12 inches) above the ground. This is usually read just once a day to give the daily rainfall total. In order to provide rainfall information through the day, a tipping-bucket rain gauge records every time 0.2 mm of rain is collected.

Another 19th century invention is the Campbell-Stokes sunshine recorder which uses a glass sphere to score a line on to special time-marked cards that can be analysed to give sunshine durations. These cards are today being replaced by data from automatic solar radiation sensors. Wind speed is measured by an anemometer, with three rotating cups, and wind direction recorded using a wind vane. These wind sensors are usually exposed at 10 m above the ground. The unit often used to measure wind speed, the knot or nautical mile per hour, is a reminder of the importance of the wind to seafarers. The list of other equipment usually includes a barometer and, if an automated site, a visibility meter, a cloud-base recorder and a snow-depth sensor.

Over the years, the Met Office has collected a vast amount of British weather data, the earliest continuous records being those from the Radcliffe Observatory at Oxford (1815) and Durham University Observatory (1843). By examining diaries and other contemporary documents, it has been possible to identify some data from even earlier periods. For example, a monthly Central England temperature series has been assembled for 1659 onwards.

The maps in this book use surface data from land-based weather stations. However, observing equipment can also be found on buoys, light vessels and ships at sea to monitor the weather off our shores and beyond. A variety of other technology is used to monitor conditions up through the atmosphere including rainfall radars, lightning detectors, aircraft and satellites. Monitoring and forecasting the weather is a truly global activity, with observing standards and the exchange of data between countries coordinated by the World Meteorological Organization – the United Nations' voice on the weather and climate.

For further information about the Met Office and the weather services it provides, see http://www.metoffice.gov.uk/

Acknowledgements

This book was authored by John Prior and edited by Sarah Tempest.
Dan Hollis and Matthew Perry produced the 1971–2000 averages maps.
Thanks also to David Sexton who was an advisor for the text regarding the future climate maps in chapter 6.

Index